Chloroacetamide Herbicides and Their Transformation Products in Drinking Water

About the Awwa Research Foundation

The Awwa Research Foundation (AwwaRF) is a member-supported, international, nonprofit organization that sponsors research to enable water utilities, public health agencies, and other professionals to provide safe and affordable drinking water to consumers.

The Foundation's mission is to advance the science of water to improve the quality of life. To achieve this mission, the Foundation sponsors studies on all aspects of drinking water, including supply and resources, treatment, monitoring and analysis, distribution, management, and health effects. Funding for research is provided primarily by subscription payments from approximately 1,000 utilities, consulting firms, and manufacturers in North America and abroad. Additional funding comes from collaborative partnerships with other national and international organizations, allowing for resources to be leveraged, expertise to be shared, and broad-based knowledge to be developed and disseminated. Government funding serves as a third source of research dollars.

From its headquarters in Denver, Colorado, the Foundation's staff directs and supports the efforts of more than 800 volunteers who serve on the board of trustees and various committees. These volunteers represent many facets of the water industry, and contribute their expertise to select and monitor research studies that benefit the entire drinking water community.

The results of research are disseminated through a number of channels, including reports, the Web site, conferences, and periodicals.

For subscribers, the Foundation serves as a cooperative program in which water suppliers unite to pool their resources. By applying Foundation research findings, these water suppliers can save substantial costs and stay on the leading edge of drinking water science and technology. Since its inception, AwwaRF has supplied the water community with more than $300 million in applied research.

More information about the Foundation and how to become a subscriber is available on the Web at **www.awwarf.org**.

Chloroacetamide Herbicides and Their Transformation Products in Drinking Water

Prepared by:
Michelle L. Hladik, **A. Lynn Roberts,** and **Edward J. Bouwer**
Johns Hopkins University
313 Ames Hall
3400 North Charles Street
Baltimore, MD 21218

Jointly sponsored by:
Awwa Research Foundation
6666 West Quincy Avenue, Denver, CO 80235-3098

and

U.S. Environmental Protection Agency
Washington, DC 20460

Published by:

DISCLAIMER

This study was jointly funded by the Awwa Research Foundation (AwwaRF) and the U.S. Environmental Protection Agency (USEPA) under Cooperative Agreement No. R829409-01. AwwaRF and the USEPA assume no responsibility for the content of the research study reported in this publication or for the opinions or statements of fact expressed in the report. The mention of trade names for commercial products does not represent or imply the approval or endorsement of AwwaRF or the USEPA. This report is presented solely for informational purposes.

Printed in the U.S.A.

Printed on recycled paper

CONTENTS

TABLES

FIGURES

FOREWORD

The Awwa Research Foundation (AwwaRF) is a nonprofit corporation that is dedicated to the implementation of a research effort to help utilities respond to regulatory requirements and traditional high-priority concerns of the industry. The research agenda is developed through a process of consultation with subscribers and drinking water professionals. Under the umbrella of a Strategic Research Plan, the Research Advisory Council prioritizes the suggested projects based upon current and future needs, applicability, and past work; the recommendations are forwarded to the Board of Trustees for final selection. The foundation also sponsors research projects through an unsolicited proposal process; the Collaborative Research, Research Applications, and Tailored Collaboration programs; and various joint research efforts with organizations such as the U.S. Environmental Protection Agency, the U.S. Bureau of Reclamation, and the Association of California Water Agencies.

This publication is a result of one of these sponsored studies, and it is hoped that its findings will be applied in communities throughout the world. The following report serves not only as a means of communicating the results of the water industry's centralized research program but also as a tool to enlist the further support of the nonmember utilities and individuals.

Projects are managed closely from their inception to the final report by the foundation's staff and large cadre of volunteers who willingly contribute their time and expertise. The foundation serves a planning and management function and awards contracts to other institutions such as water utilities, universities, and engineering firms. The funding for this research effort comes primarily from the Subscription Program, through which water utilities subscribe to the research program and make an annual payment proportionate to the volume of water they deliver and consultants and manufacturers subscribe based on their annual billings. The program offers a cost-effective and fair method for funding research in the public interest.

A broad spectrum of water supply issues is addressed by the foundation's research agenda: resources, treatment and operations, distribution and storage, water quality and analysis, toxicology, economics, and management. The ultimate purpose of the coordinated effort is to assist water suppliers to provide the highest possible quality of water economically and reliably. The true benefits are realized when the results are implemented at the utility level. The foundation's trustees are pleased to offer this publication as a contribution toward that end.

Walter J. Bishop
Chair, Board of Trustees
Awwa Research Foundation

Robert C. Renner, P.E.
Executive Director
Awwa Research Foundation

ACKNOWLEDGMENTS

The authors of this report are indebted to the following water utilities and individuals for their cooperation and participation in this project:

American Water, Voorhees, NJ, Mark LeChevallier
City of Columbia Water, Columbia, MO, John Betz
Louisville Water Company, Louisville, KY, Rengao Song

In addition, the advice of the Project Advisory Committee (PAC)—including Jody Shoemaker, National Exposure Research Laboratory, U.S. Environmental Protection Agency, Cincinnati, OH—and help of the Awwa Research Foundation project officer, Djanette Khiari, are appreciated.

The authors wish to acknowledge Jonie Hsiao, Dan Carlson, Khoi Than and Mike Blumenfeld for their help in the synthesis of the neutral chloroacetamide degradates.

EXECUTIVE SUMMARY

The chloroacetamide herbicides (acetochlor, alachlor, metolachlor and dimethenamid) and the triazines (atrazine, cyanazine and simazine) are two of the most widely used classes of herbicide in the United States. This class of herbicides is applied mainly in spring, and is used on agricultural crops (especially corn, soybeans and sorghum). The highest intensity of application for these herbicides is in the Midwestern United States. These herbicides are frequently encountered in groundwater, surface water and drinking water in areas of high usage.

While the parent herbicides have been well documented in natural waters, relatively little is known about the occurrence of herbicide degradates in water. Those chloroacetamide degradates (the ionic ethane sulfonic acids and oxanilic acids) that have been previously studied in natural waters are found more frequently and at higher concentrations than the parent herbicide. The chloroacetamide herbicides are also known to degrade to many other neutral degradates that have not been studied extensively in natural waters. Two atrazine degradates (desethyl atrazine and desisopropyl atrazine) have been extensively studied and typically occur at concentrations less than atrazine.

The parent herbicides are known to have adverse human health effects; typically alachlor, atrazine, and simazine are regulated in drinking water. The ionic degradates are thought to be relatively nontoxic and therefore have not received much attention in drinking water despite their higher detection frequencies and concentrations. In contrast, the neutral degradates of chloroacetamides and triazines may possess a toxicity similar to the parent herbicide.

RESEARCH OBJECTIVES

The Awwa Research Foundation (AwwaRF) funded this study to determine the occurrence and removal of neutral chloroacetamide degradates in drinking water sources. This report presents the results of the study in terms of occurrence of neutral chloroacetamide degradates in drinking water sources (groundwater and surface water), their removal in finished drinking water, and their removal in bench scale unit processes. The key objectives of the study included the following:

1. Determine neutral chloroacetamide degradates to target for analysis through a search of the literature.
2. Synthesize the target neutral chloroacetamide degradates to provide analytical standards.
3. Develop a method for the concentration and analysis of low-level concentrations of triazines, chloroacetamides and neutral chloroacetamide degradates in drinking water using solid-phase extraction (SPE) and gas chromatography/mass spectrometry (GC/MS).
4. Quantify the neutral degradates in raw and treated drinking water from a series of Midwestern drinking water utilities at two time periods, spring and fall.
5. Quantify the ionic degradates in raw and treated drinking water using previously developed methods from a series of Midwestern drinking water utilities at two time periods, spring and fall.
6. Conduct bench scale unit processes simulating drinking water treatment to ascertain which methods might be most effective for removing these compounds.

APPROACH

An initial literature review was conducted before experiments were begun to determine which neutral chloroacetamide degradates would most likely be found in drinking water supplies.

Synthesis protocols for the selected neutral degradates were established by using either previously developed methods or methods developed in our laboratory. Authentic standards of all compounds under investigation allowed for the water analysis method development stages to commence.

A method for the concentration of the neutral degradates (plus chloroacetamides and triazines) from water was achieved using SPE, including water that contained preservatives used in the shipping of samples from the utilities. The extracts were quantified using GC/MS. Sites for drinking water samples were chosen to focus on areas of high chloroacetamide herbicide use. A broad cross-section of source water and treatment trains was also included in finalizing site selection.

Once an analytical method was validated the neutral and ionic degradates could be measured in water samples. The first round of raw and treated drinking water sampling was completed in Fall 2003, and the second set was completed in the Spring 2004. Bench scale unit processes using an amended surface water were completed to complement the analysis of drinking water utility samples.

CONCLUSIONS

1. Through a literature search, 20 neutral chloroacetamide degradates were chosen for the study. Six ionic degradates were also included.
2. A robust method was developed for the analysis of neutral chloroacetamide herbicides (plus chloroacetamides and triazines) at the pg/L to ng/L range using SPE and GC/MS.
3. Preservatives were identified that were suitable for addition to drinking water samples undergoing shipment. There preservatives had no discernible effects on recovery of the neutral degradates or parent herbicides. Preserved samples containing neutral degradates could be stored in a refrigerator for up to one month without significant loss of analytes.
4. The samples for the ionic degradates did not contain preservatives because of interferences they introduced in SPE recoveries.
5. Neutral chloroacetamide herbicide degradates were encountered in raw and finished drinking water in fall and spring samples. Concentrations were slightly higher in the spring although overall concentrations were similar (~10-100 ng/L).
6. The parent chloroacetamide herbicides were encountered in both fall and spring, with spring concentrations being approximately one order of magnitude higher (13-300 ng/L) for acetochlor, metolachlor and dimethenamid than in fall (3-11 ng/L). Median concentrations of alachlor were lower in the spring (3 ng/L) than in the previous fall (7 ng/L).
7. Ionic degradate concentrations remained fairly constant during both sampling periods (~100-1000 ng/L).
8. The parent triazines and neutral degradates were found at higher concentrations in the spring (10-1600 ng/L) than in the fall (7-80 ng/L).
9. No drinking water utilities experienced substantial removals of the parent herbicides or degradates in the fall.

10. Drinking water utilities that employed activated carbon (granular or powdered) encountered significant removals in the spring (average removal for all compounds ~40%).
11. The site that employed ozone encountered significant removals of some compounds, although other degradates appear to have been produced upon treatment.
12. Those sites that employed coagulation/flocculation, sand filtration and chlorination exhibited no significant overall removals of the target analytes.
13. There was no discernible trend between concentrations in the groundwater and surface water sample concentrations.
14. Bench scale studies showed that coagulation with alum (30 mg/L) and ferric chloride (20 mg/L) was ineffective at removing the neutral chloroacetamide degradates or the parent chloroacetamide herbicides.
15. Chlorination (6 mg/L for 6 hrs) proved effective at removing those compounds that lack the acetanilide functional group.
16. Ozonation (3 mg/L for 30 min) proved effective at removing all compounds.
17. All compounds were amenable to adsorption onto activated carbon. Those compounds that have a lower estimated K_{ow} were found to have a lower adsorption affinity for the activated carbon.

RECOMMENDATIONS

1. If the neutral chloroacetamide degradates were to be regulated in the future, this could necessitate removal through additional treatment.
2. If the neutral chloroacetamide degradates were to be regulated in the future in could effect removal, albeit at additional cost. Ozone proved effective in removing the parent herbicides, although it may not be possible to achieve complete mineralization using doses and contact times typically employed during drinking water treatment.
3. Utilities located in regions of heavy herbicide use should monitor herbicides and degradates proactively, increasing the knowledge about these degradates.

FUTURE RESEARCH

Whether the neutral chloroacetamide degradates are regulated in drinking water may depend on their toxicity to humans. Currently there are insufficient data to provide a definitive assessment as to whether these neutral chloroacetamide degradates are a human health concern.

If neutral chloroacetamide degradates were to be regulated, then the most cost effective method for removing the degradates (without creating new ones) would have to be determined. Our research suggests activated carbon is an option, although this may be associated with increased costs to utilities if removal of the more water soluble degradates is required. Ozone, at sufficient doses, may prove effective for all of the degradates under investigation, although this too may be associated with additional costs.

Additional experiments regarding drinking water treatment should be conducted as with respected to reaction kinetics, and the influence of oxidant dosage and contact time. Other experiments should be performed to determine the behavior of these compounds with membrane treatment and their potential biodegradation.

Our study found no significant difference in concentrations of neutral chloroacetamides between groundwater and surface water sources at some of the sites sampled. Although it is often assumed that pesticide residues are present at lower concentrations in groundwater than in surface water, such a tendency was not encountered in our studies. Additional studies would be required to further explore this question.

CHAPTER 1
INTRODUCTION

OCCURRENCE OF HERBICIDES IN DRINKING WATER

The chloro-*s*-triazine and chloroacetamide herbicides represent the two most extensively employed (Table 1.1) groups of herbicides in the U.S. (each responsible for 20% of conventional agricultural pesticide use in 1997 in terms of weight of active ingredient, and often administered simultaneously; Aspelin and Grube 1999). They are mainly used for pre-emergence control of annual grasses and broadleaf weeds, primarily on corn, sorghum, and soybeans. Their use is thus heavily concentrated in the Midwestern U.S., as illustrated in Figures 1.1 and 1.2 for atrazine and metolachlor. Other herbicides in these classes follow similar usage patterns, although simazine is most extensively used on fruit crops.

Herbicides within these two classes exhibit modest hydrophobicity (log K_{ow} values; Table 1.1) and are thus readily leached from agricultural soils into receiving ground and surface waters. They are at best slowly degraded under the dilute conditions characteristic of natural ground waters and surface waters, and form relatively persistent degradates.

As might have been anticipated from their modest hydrophobicity, moderate persistence, and extensive use, recent studies conducted by the USGS have revealed the widespread occurrence of these herbicides in drinking water sources. For example, they have been detected in 90-99% of Midwestern surface water samples analyzed (Table 1.2) (Battaglin et al. 2000). Concentrations in surface water are often high relative to drinking water standards; the median concentration of atrazine determined in 129 Midwestern surface water samples, for example, exceeded the MCL value of 3 μg/L, while the maximum alachlor concentration was nearly an order of magnitude in excess of its MCL value. These two classes are among the most frequently detected pesticides in groundwater (Kolpin, Barbash and Gilliom 2000), and groundwater concentrations tend to be 1-2 orders of magnitude lower than in surface water (Kolpin, Barbash and Gilliom 1998).

OCCURRENCE OF HERBICIDE DEGRADATES IN DRINKING WATER

One of the more troubling outcomes of the USGS findings relates to the prevalence of herbicide degradates in surface and groundwater. Chloroacetamide and chloro-*s*-triazine degradates have been found in up to 60% of Iowa municipal wells sampled during 1996 (Kolpin, Thurman and Linhart 1998). Several such herbicide degradates are found at concentrations equivalent to or even in excess of that of the parents (Kolpin, Thurman and Linhart 1998).

We hypothesize that those herbicide degradates that have been measured to date in drinking water sources represent the mere "tip of the iceberg." As any analytical chemist knows, "what you see depends on what you look for," and this statement is particularly applicable to the analysis of organic micropollutants in water. Dealkylated triazines have proved relatively easy to quantitate in environmental samples, in large part because authentic standards were available. This has contributed to their popularity as analytes. Other herbicide degradates, for which reference materials are not commercially available, have received much less attention – not necessarily because they are of lesser concern, but simply because their quantitation is more difficult.

Table 1.1
Important chloroacetamide and chloro-*s*-triazine pesticides in U.S. agriculture

Herbicide	Structure	Class	1997 U.S. ranking and total use*	log K_{ow}†	Regulatory considerations
Atrazine	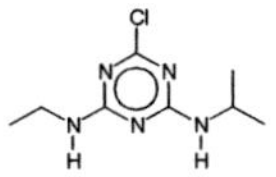	Chloro-*s*-triazine	# 1 75-82 mil. lbs. AI	2.61	MCL 3 µg/L; MCLG 3 µg/L‡
Metolachlor	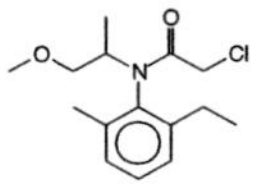	Chloro-acetamide	# 2 63-69 mil. lbs. AI	3.13	EPA CCL contaminant§
Acetochlor	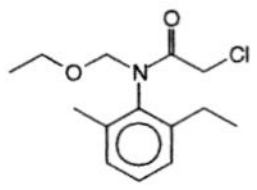	Chloro-acetamide	# 7 31-36 mil. lbs. AI	2.39	EPA CCL contaminant§
Cyanazine	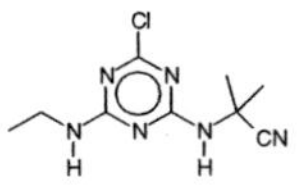	Chloro-*s*-triazine	# 11 18-22 mil. lbs. AI	2.22	EPA CCL contaminant§
Alachlor	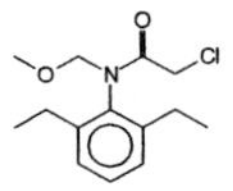	Chloro-acetamide	#12 13-16 mil. lbs. AI	3.52	MCL 2 µg/L; MCLG zero‡
Dimethenamid	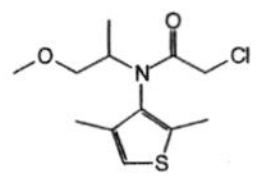	Chloro-acetamide	#20 6-9 mil. lbs. AI	1.57	—
Simazine	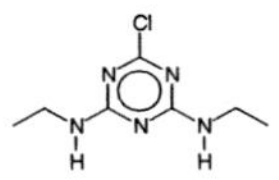	Chloro-*s*-triazine	#23 5-7 mil. lbs. AI	2.18	MCL 4 µg/L; MCLG 4 µg/L‡

* Data from Aspelin and Grube 1999
† Biobyte ClogP
‡ Data from EPA 2001b
§ Data from EPA 1998b

A telling example is that of the ESA degradates of chloroacetamides. For many years, their existence was overlooked. Researchers suspected something was "missing" when a comparison of GC/MS analyses for chloroacetamides to those obtained using enzyme-linked immunoassay (ELISA) methods indicated a positive bias in the latter (Baker et al. 1993). This set a flurry of "analytical detective" activity (Barrett 1996; Field and Thurman 1996) into motion. Early literature pertaining to ELISA method development had noted significant cross-reactivity in ESA degradates (Feng 1991); on developing the requisite methods and reanalyzing samples, researchers confirmed the false positives could be attributed to ESA degradates (Barrett 1996).

Follow-up studies (mostly, but not exclusively, conducted by USGS researchers) revealed the ubiquity of ESA degradates in surface and groundwater (Aga et al. 1996; Kolpin, Thurman and Goolsby 1996; Thurman et al. 1996; Ferrer, Thurman and Barceló 1997; Kalkhoff et al. 1998; Kolpin, Thurman and Linhart 1998; Graham et al. 1999; Phillips et al. 1999; Boyd 2000; Kolpin,

ATRAZINE

ESTIMATED ANNUAL AGRICULTURAL USE

Average use of Active Ingredient
Pounds per square mile of county per year

- No Estimated Use
- < 1.080
- 1.080 – 5.587
- 5.588 – 21.211
- 21.212 – 66.515
- >= 66.516

Crops	Total Pounds Applied	Percent National Use
corn	53,796,206	84.13
sorghum	7,339,963	11.48
sugar cane: sugar & seed	1,711,322	2.68
pasture	518,074	0.81
sweet corn	444,523	0.70
sod	119,182	0.19
prosomillet	11,937	0.02
field and grass seed	6,305	0.01

Source: USGS 1998

Figure 1.1 Atrazine use in U.S., 1990-1995

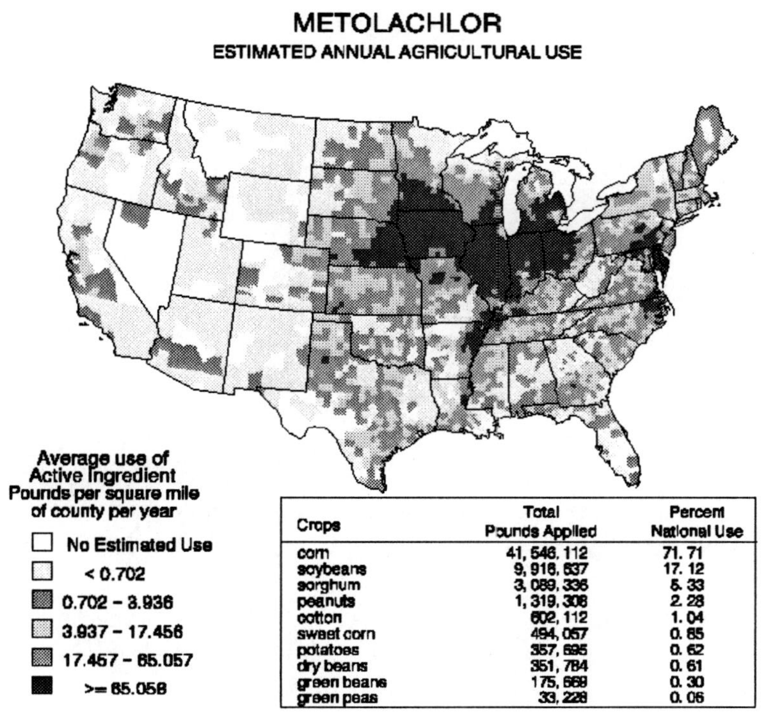

Crops	Total Pounds Applied	Percent National Use
corn	41,546,112	71.71
soybeans	9,916,637	17.12
sorghum	3,089,336	5.33
peanuts	1,319,308	2.28
cotton	602,112	1.04
sweet corn	494,057	0.85
potatoes	357,595	0.62
dry beans	351,784	0.61
green beans	175,669	0.30
green peas	33,228	0.06

Source: USGS 1998

Figure 1.2 Metolachlor use in U.S., 1990-1995

Table 1.2
Summary of chloroacetamide and chloro-*s*-triazine herbicide concentrations determined by USGS in 129 water samples from Midwestern streams and rivers in 1998

Herbicide	Detections above minimum reporting level	Median concentration (μg/L)	Maximum concentration (μg/L)
Acetochlor	124	0.411	25.1
Alachlor	116	0.045	17.2
Atrazine	129	3.97	224
Cyanazine	119	0.326	14.0
Metolachlor	129	1.73	143

Source: Data from Battaglin et al. 2000

Table 1.3
Concentrations of chloroacetamide ESA and OA degradates in ARP groundwater and finished surface water studies (see Table 1.4 for structures)

Analyte	95th percentile concentration in groundwater (2nd quarter, 1999) (μg/L)	Maximum concentration in finished surface water (μg/L)	95th percentile concentration in finished surface water (μg/L)
Acetochlor-ESA	2.59	0.7	< 0.50
Acetochlor-OA	< 0.50	1.26	0.56
Alachlor-ESA	3.95	1.75	< 0.50
Alachlor-OA	< 0.50	0.57	< 0.50
Metolachlor-ESA	6.59	1.67	1.01
Metolachlor-OA	0.78	0.92	0.66

Source: Data from ARP 2002

Thurman and Linhart 2000). Analyses of 175 wells conducted by the Acetochlor Registration Partnership (ARP) during 1999-2001 confirm these findings for groundwater from "corn belt" states (USGS, 1998). ARP studies have also revealed the presence of the acetochlor ESA and OA Degradates in 20 and 36%, respectively, of finished drinking water samples derived from 175 surface water sources (Table 1.3).

The ESA and OA derivatives are not the only chloroacetamide degradates likely to be more prevalent than the parent herbicides. For example, one study reports the alachlor degradate 2,6-diethylaniline in 16% of Midwestern groundwater samples (as opposed to a 3.3% detection frequency for alachlor) (Kolpin, Thurman and Goolsby 1996). A much broader range of chloroacetamide degradates has been tentatively identified (through mass spectral interpretative techniques, occasionally confirmed with authentic standards) to result from biological or abiotic transformations of chloroacetamides in laboratory settings. For example, more than 33 degradates have been reported to result from degradation of alachlor and metolachlor, respectively (Chesters et al. 1989). Information pertaining to the most important chloroacetamide degradates is summarized in Table 1.4. Studies submitted to EPA to support registration imply that non-ESA degradates may be more prevalent than the ESA degradates (Barrett 1996). Many chloroacetamide degradates are less hydrophobic than the parent herbicides (Table 1.4), and are considerably more mobile and are also of at least moderate persistence (Barrett 1996). Certain herbicide degradates (such as dealkylated chloro-*s*-triazines, hydroxyalachlor, and deschloroalachlor) can actually be formed during water treatment (*e.g.*, during ozonation; Acero, Stemmler and von Gunten 2000, or UV irradiation; Somich et al. 1988). Some are known to evince substantial toxicity (see below). That few chloroacetamide degradates (other than the ESA and OA degradates) have been measured in environmental samples in large part reflects the unavailability of reference materials needed for quantitative analysis, and does not signify an inherently low risk to human health.

Apart from the USGS's and ARP's analyses of chloroacetamide ESA and OA derivatives, the most noteworthy prior attempt to measure chloroacetamide degradates in environmental samples was published by Potter and Carpenter (1995). These researchers synthesized reference materials for 10 different alachlor degradates, and sought them in groundwater samples collected beneath a Massachusetts corn field. Six of these degradates were identified, along with 14 other compounds whose electron ionization (EI) and positive chemical ionization (pCI) mass spectra indicated they were related to alachlor. Concentrations of individual alachlor degradates for which reference materials were available ranged from 4 to 570 ng/L, and total degradate concentrations exceeded that of the parent compound (370-1,100 ng/L) by a factor in excess of 2. One limitation of this prior study was that it employed liquid/liquid extraction methods for concentrating analytes. Although this method can work well for highly hydrophobic analytes, it is less well suited to more polar analytes (such as herbicide degradates). Indeed, reported recoveries for alachlor, 2,6-diethylaniline, and anthracene-d_{10} in deionized water only ranged between 12-20% (Potter and Carpenter 1995). Although no information was provided as to recoveries for herbicide degradates (other than 2,6-diethylaniline), it is safe to assume that these were low and that many relatively polar degradates might well have been missed altogether. Alternative concentration methods (such as solid phase extraction (SPE) techniques employing media with high affinities for polar analytes) might well reveal the presence of herbicide degradates not detected by Potter's work.

Although alachlor was formerly one of the most extensively used of the chloroacetamides, and still remains in many hydrologic systems, its usage has diminished sharply in recent years, owing to its replacement by acetochlor (registered for use in the U. S. in 1994). Environmental degradates of metolachlor and acetochlor may be equally likely to occur in shallow groundwater and surface water used as drinking water sources. To date, apart from USGS and ARP studies of the ESA and OA chloroacetamide derivatives and some measurements of 2,6-diethylaniline, comprehensive studies of chloroacetamide degradates in drinking water sources are lacking.

Many different chloroacetamide degradates that have rarely if ever been sought can reasonably be expected to be present in drinking water sources. Some, as discussed below, are of concern by virtue of their known toxicity or mutagenicity (or ability to undergo metabolism to potent mutagens).

Table 1.4
Examples of important chloroacetamide degradates. Note that all are EPA CCL contaminants. Unless otherwise indicated, all degradates were identified from lab studies only.

Transformation product	Structure	Occurrence and *toxicity*
Metolachlor Degradates		
4-(2-Ethyl-6-methylphenyl)-5-methyl-3-morpholinone	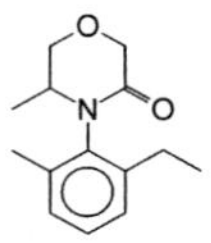	Major (74-84%) photolysis product (Mathew and Kahn 1996), and important microbial degradate (Liu et al. 1989). Identified in soils after application of metolachlor (Chesters et al. 1989).
Hydroxymetolachlor	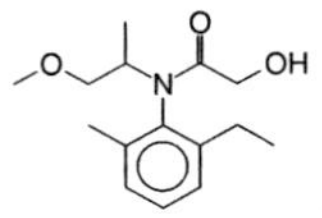	Important fungal and bacterial degradate (Chesters et al. 1989; Liu, Freyer and Bollag 1991; Sanyal and Kulshrestha 2002); major product of basic hydrolysis (Carlson et al. 2003); minor photoproduct (Kochany and Maguire 1994; Mathew and Kahn 1996). Identified in soils after application of metolachlor (Chesters et al. 1989).
Deschloro metolachlor	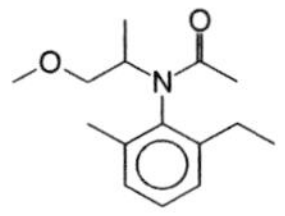	Identified as degradate in soil:water and sediment:water slurries (Chesters et al. 1989) and as photoproduct (Kochany and Maguire 1994). Major product of metolachlor reaction with FeS_2 (Carlson 2003).
Deschloroacetylmetolachlor	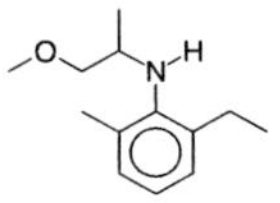	Inferred photoproduct (5-30%) (Wilson and Mabury 2000). Minor product of metolachlor reaction with FeS_2 (Carlson 2003).
2-Chloro-2'-ethyl-6'-methylacetanilide	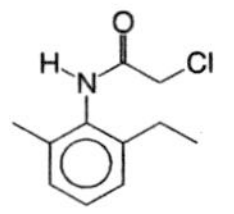	Identified as fungal (Sanyal and Kulshrestha 2002) and as possible bacterial (Liu, Freyer and Bollag 1991) metabolite. Note that this product could also result from degradation of acetochlor; note also *mutagenicity* (Tessier and Clark 1995) of analagous alachlor degradate.
2-Hydroxy-2'-ethyl-6'-methylacetanilide	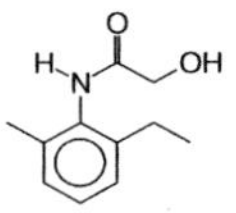	Photolysis product (Kochany and Maguire 1994). Note that this degradate could also result from degradation of acetochlor; note also *mutagenicity* (Tessier and Clark 1995) of analagous alachlor degradate.
2-Chloro-2'-ethyl-6'-methyl-*N*-(2-hydroxy-1-methylethyl)-acetanilide	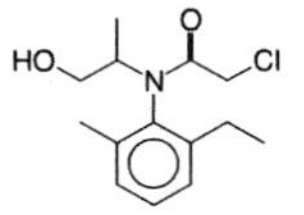	Major bacterial (Liu, Freyer and Bollag 1991) and fungal (Pothuluri et al. 1997) metabolite. Identified in soils after application of metolachlor (Chesters et al. 1989).
Metolachlor ESA (ethane sulfonic acid)	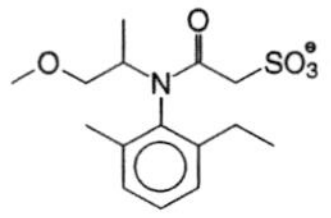	Widely encountered at relatively high concentrations in groundwater and surface water.
Metolachlor OA (oxanilic acid)	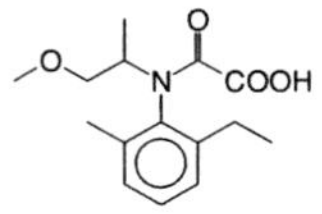	Widely encountered at relatively high concentrations in groundwater and surface water.

(continued)

Table 1.4 (Continued)

Transformation product	Structure	Occurrence and *toxicity*
2-Ethyl-6-methyl aniline	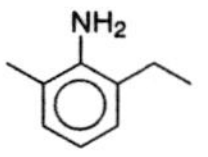	Fungal metabolite of metolachlor (Sanyal and Kulshrestha 2002). Formed as end product of acid-catalyzed hydrolysis of metolachlor and acetochlor (Carlson 2003). *Promutagen* (Kimmel, Casida and Ruzo 1986).
Alachlor Degradates		
Hydroxyalachlor	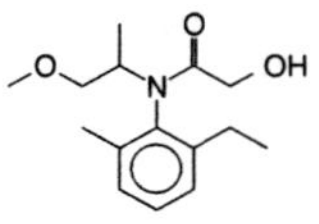	ID'd as hydrolysis product (Carlson 2003); as microbial degradate (Chesters et al. 1989); as degradate in soil/sediment:water slurries (Chesters et al. 1989); and as photoproduct (Somich et al. 1988; Chiron et al. 1995; Hogenboom, Niessen and Brinkman 2000). Found (up to 100 ng/L) in groundwater (Potter and Carpenter 1995). *More toxic than parent* (Kross, Vergara and Raue 1992).
Deschloroalachlor	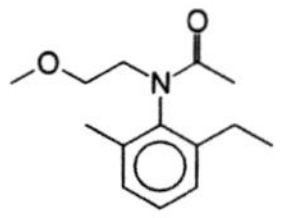	ID'd as degradate in soil/sediment:water slurries (Chesters et al. 1989); as photoproduct (Somich et al. 1988); and as product of reduction by FeS_2 (Carlson 2003). Found (100-550 ng/L) in groundwater (Potter and Carpenter 1995).
2,6-Diethylacetanilide	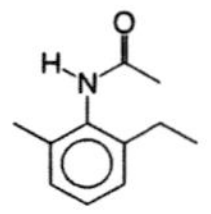	Photoproduct (Somich et al. 1988; Chiron et al. 1995); degradate in soil:water slurries (Chesters et al. 1989). Found (up to 130 ng/L) in groundwater (Potter and Carpenter 1995). Potential diethylaniline DBP (Hwang, Larson and Snoeyink 1990).
2-Chloro-2',6'-diethylacetanilide	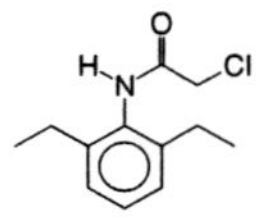	Fungal, bacterial and photolytic degradate (Chesters et al. 1989). Measured in river water and groundwater (Pereira, Rostad and Leiker 1990). Sought but not detected by Potter and Carpenter (1995). *Mutagenic* (Tessier and Clark 1995).
2-Hydroxy-2',6'-diethylacetanilide	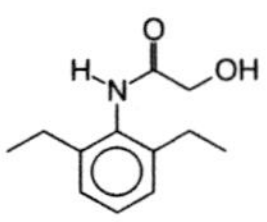	Fungal metabolite (Chesters et al. 1989). Measured in river water and groundwater (Pereira, Rostad and Leiker 1990). Sought but not detected by Potter and Carpenter (1995). *Mutagenic* (Tessier and Clark 1995) .
2-Hydroxy-2',6'-diethyl-*N*-methyl acetanilide	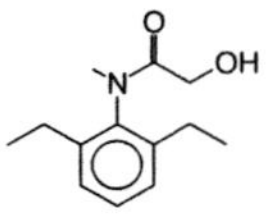	Minor photoproduct (Somich et al. 1988; Peñeula and Barceló 1996; Hogenboom, Niessen and Brinkman 2000). Found by Potter and Carpenter (1995) at up to 130 ng/L in MA groundwater.
Alachlor ESA	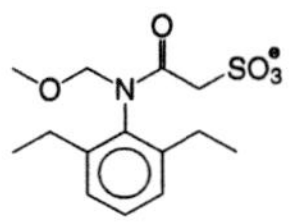	Widely encountered at relatively high concentrations in groundwater and surface water.
Alachlor OA	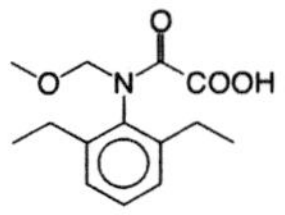	Widely encountered at relatively high concentrations in groundwater and surface water.

(continued)

Table 1.4 (Continued)

Transformation product	Structure	Occurrence and *toxicity*
2,6-Diethylaniline	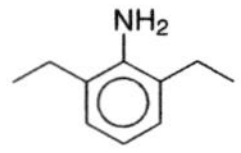	Fungal and mammalian degradate and photoproduct (Chesters et al. 1989). Found (up to 85 ng/L) in nationwide survey (Kolpin, Thurman and Linhart 1998). Found (5-16 ng/L) in groundwater (Potter and Carpenter 1995) and surface water (13-95 ng/L) (Galassi et al. 1996). Minor product of reaction with FeS_2 (Carlson 2003). *Promutagen* (Kimmel, Casida and Ruzo 1986).
Acetochlor Degradates		
Hydroxyacetochlor	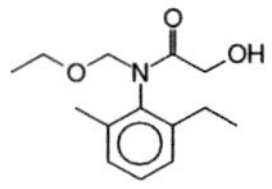	Formed during basic hydrolysis in lab (Carlson 2003); likely product of bacterial degradation of acetochlor ESA and of acetochlor photolysis.
Deschloroacetochlor	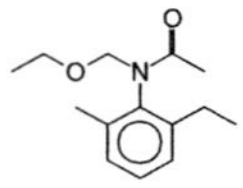	Likely product of reduction of acetochlor by redox-active mineral phases such as FeS_2.
2-Ethyl-6-methylacetanilide	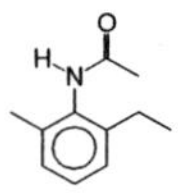	Potential photoproduct. (Note this can also originate from metolachlor.)
Acetochlor ESA	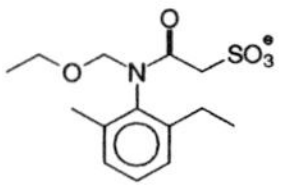	Widely encountered at relatively high concentrations in groundwater and surface water.
Acetochlor OA	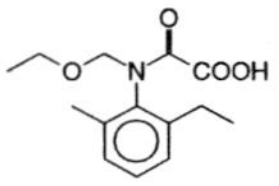	Widely encountered at relatively high concentrations in groundwater and surface water.

Most degradates can be formed by more than one route. For example, we find that the amide cleavage product deschloroacetylmetolachlor, previously proposed to result from photolysis of metolachlor (Wilson and Mabury 2000), is also generated as a minor product of the reaction of metolachlor with iron pyrite in deoxygenated laboratory solution (Carlson 2003). This latter reaction may reflect catalysis of amide bond cleavage by a Lewis acid, such as a metal oxide. Because of the prevalence of metal oxides in aerobic subsurface environments, such dealkylated degradates may be much more prevalent than is currently recognized. Our field results (Hladik, Hsiao and Roberts 2005) in a vertical profile obtained from the Chesapeake Bay indicate that its concentrations increase markedly with depth, implying that it does not originate through photolysis in the Chesapeake Bay, but rather may be forming within the water column (perhaps through biological transformations) or else is introduced by groundwater inflow. Other examples of chloroacetamide degradates that can form from multiple routes include the reductive dechlorination products (which can result from photolysis (Somich et al. 1988; Kochany and Maguire

1994), from biodegradation (Chesters et al. 1989), and through reduction by minerals (Carlson 2003), and the hydroxy-substituted chloroacetamide degradates (which can form from hydrolysis; Carlson 2003), photolysis; Somich et al. 1988; Peñeula and Barceló 1996; Zheng and Ye 2001; Christen 2002, or biodegradation; Chesters et al. 1989). Notably, they have also been shown to represent the major microbial metabolites of ESA degradates of chloroacetamides (Laue, Field and Cook 1996). The hydroxy-substituted product of alachlor is more toxic than the parent compound (Kross, Vergara and Raue 1992). Because of the prevalence of the ESA degradates, these hydroxy-substituted acetamides could also represent toxicologically important degradates whose importance has been overlooked in prior analyses.

Our laboratory studies show that abiotic *N*-dealkylation results from acid-catalyzed hydrolysis of alkoxymethyl-substituted chloroacetamides (e.g., alachlor, acetochlor, butachlor; Carlson 2003). If such reactions were catalyzed by metal oxide surfaces, they would give rise to the mutagenic (Tessier and Clark 1995) alachlor degradate 2-chloro-2',6'-diethylacetanilide, previously reported to result from alachlor biodegradation and photolysis (Chesters et al. 1989). An analagous reaction for acetochlor would give rise to 2-chloro-2'-ethyl-6'-methylacetanilide, previously reported as a metolachlor metabolite (Liu, Freyer and Bollag 1991; Sanyal and Kulshrestha 2002).

A wide variety of different chloroacetamide degradates may hence occur in raw drinking water sources. Whether they are efficiently removed during conventional treatment remains to be determined. The toxicity of at least some of these degradates raises concerns, as discussed below.

HEALTH CONSEQUENCES ASSOCIATED WITH HERBICIDES IN DRINKING WATER

The presence of herbicides and their degradates in drinking water sources poses questions directly related to human health. Atrazine, although classified as "not likely" to be carcinogenic to humans (EPA 1999), is a potent endocrine disrupter (Hayes et al. 2002) that has been shown to disturb hypothalamic-pituitary function in laboratory animals, and it is not unreasonable to expect similar effects in humans (EPA 2002a). Potential consequences to humans include reproductive disruption in females and prostatitis in males (EPA 2002a). Simazine, closely related in structure to atrazine, is believed to share a common mechanism of toxicity (EPA 2002b). Alachlor is viewed as a "likely" (class B-2) human carcinogen (EPA 1998a) and a potential teratogen (Osano, Admiral and Otieno 2002). Industry-sponsored studies (Heydens et al. 1999) claim it exerts its carcinogenicity through non-genotoxic mechanisms, suggesting that animal studies may have limited applicability to humans and that a carcinogenicity threshold may exist. This finding is disputed by EPA analyses of mutagenicity and carcinogenicity data provided by applicants seeking to register chloroacetamides with the Office of Pesticide Programs. In fact, the EPA analyses reveal a pattern of clastogenicity that is consistent with genotoxic activity (Dearfield et al. 1999). Such activity may be related to the alkylating ability of the chloroacetamide group (Dearfield et al. 1999); this could lead to binding to DNA as well as to depletion of glutathione (GSH) levels, the latter leading to increased susceptibility to damage to tissues with low endogenous levels of GSH (Dearfield et al. 1999). Alternatively, the carcinogenicity of chloroacetamides may stem from release of formaldehyde or other aldehydes during *N*-dealkylation (Dearfield et al. 1999), or from generation of primary anilines (which in turn can be converted to genotoxic

2,6-dialkylbenzoquinoneimines (Hill et al. 1997) or to nitrosobenzenes; Kimmel, Casida and Ruzo 1986). Consumption of alachlor is also associated with eye, liver, kidney or spleen problems and anemia (EPA 2001b). Such toxicity and carcinogenicity concerns have led MCL levels (Table 1.1) to be established for these three herbicides in drinking water (EPA 2001b).

Acetochlor (a CCL contaminant) has been classified by the EPA as a probable (class B-2) human carcinogen (Kolpin et al. 1996, EPA 2001b). This herbicide is believed to have a toxicity profile similar to that of alachlor (Dearfield et al. 1999). Its recent conditional registration included an extensive stewardship program that includes restriction of its sale in areas vulnerable to contamination. Available evidence suggests that metolachlor (in large part because of the greater stability conferred by its relatively bulky alkoxyethyl substituent) is less of a genetic and/or carcinogenic hazard than alachlor or acetochlor (Dearfield et al. 1999). Metolachlor has thus been classified as a "Group C" possible human carcinogen, based on increases in liver tumors in rats (EPA 1995). Metolachlor also exhibits some evidence of developmental toxicity in some (but not all) test animals (EPA 1995), and is regarded as an endocrine disrupter (Rakitsky, Koblyakov and Turusov 2000). Metolachlor has also been included as an EPA CCL contaminant. Cyanazine, another CCL contaminant, was voluntarily removed from the market at the end of 1999 (EPA 2000), although residues of this herbicide (as well as its degradates, typically present at higher concentrations than the parent compound) still remain in the environment (Ferrer, Thurman and Barceló 2000; Kolpin, Thurman and Linhart 2001).

HEALTH CONSEQUENCE ASSOCIATED WITH HERBICIDE DEGRADATES IN DRINKING WATER

Unfortunately, partial degradation of herbicides does not necessarily reduce their toxicity. For example, the dealkylated chloro-*s*-triazine degradates exhibit toxicological properties that are similar to the parent herbicides, with which they share a common mode of action (EPA 2002a). Indeed, the toxicity of chloroacetamides may result at least in part from some of the metabolites (Dearfield et al. 1999).

The possibility that the carcinogenic effect of chloroacetamides results from the tendency of the chloroamide group to serve as an alkylating agent suggests that chlorinated degradates may be at least as carcinogenic as the parents. Certainly 2-chloro-2',6'-diethylacetanilide has been shown to be mutagenic (Tessier and Clark 1995; Hill et al. 1997), and it also binds to DNA (Nelson and Ross 1998). This implies that the *N*-dealkylation product of metolachlor, 2-chloro-2'-ethyl-6'-methylacetanilide, could possess some mutagenicity. In this respect we note that the common chloroacetamide with the greatest steric hindrance, metolachlor, evinces the lowest reactivity towards sulfur nucleophiles (Loch et al. 2002). That metolachlor is viewed as having a lesser carcinogenic potential than alachlor or acetochlor may reflect its increased hindrance towards alkylation of base pairs on DNA. *N*-Dealkylated degradates that retain the chloroacetamide moiety would exhibit lesser steric hindrance towards nucleophilic attack than the parent compounds, potentially conferring greater mutagenicity.

Nor does dechlorination necessarily reduce toxicity. Several products resulting from partial or complete *N*-dealkylation of chloroacetamides have been shown to be mutagenic or to undergo further metabolism to mutagens. These include the alachlor biodegradation product 2-hydroxy-2',6'-diethylacetanilide, which has been shown by the Ames test to be mutagenic (Tessier and Clark 1995). Both 2,6-diethylaniline and 2-ethyl-6-methylaniline (the latter resulting from partial biodegradation of metolachlor) are promutagens, as they undergo metabolism to 2,6-dialkylnitrosobenzenes, which have been shown to possess high direct mutagenicity in the Ames assay (Kimmel, Casida and Ruzo 1986):

alachlor (mutagenic) → 2-chloro-2',6'-diethylacetanilide (mutagenic) + 2-hydroxy-2',6'-diethylacetanilide (mutagenic) + 2,6-diethylaniline (promutagen)

Other aniline metabolites (such as 2,6-dialkylbenzoquinoneimines) are viewed as genotoxic, with 2,6-diethylbenzoquinoneimine (potentially originating from alachlor) more so than 2-ethyl-6-methylbenzoquinoneimine (which could originate either from metolachlor or from acetochlor) (Hill et al. 1997). 2,6-Diethylaniline and 2-ethyl-6-methylaniline are also teratogenic, with the latter compound being substantially more so. The biological activity of substituted anilines and their metabolites explains why metolachlor is cited as an example of a nonteratogenic herbicide that upon degradation loses toxicity but gains teratogenicity (Osano, Admiral and Otieno 2002).

Products in which the alkoxyalkyl substituent remains intact also may possess toxicity. For example, studies (Kross, Vergara and Raue 1992) conducted using the Microtox assay revealed that hydroxyalachlor possesses toxicity greater than the parent compound, although this particular degradate was not found to be mutagenic according to the Ames test (Tessier and Clark 1995). Any chloroacetamide degradate retaining an alkoxyalkyl substituent could generate formaldehyde or other reactive aldehyde intermediates (Dearfield et al. 1999).

Not all herbicide degradates are likely to represent a human health risk. The alachlor ESA degradate, for example, is poorly absorbed and undergoes minor metabolism, in contrast to the significant absorption and substantial metabolism observed with alachlor (Heydens et al. 2000). It also demonstrates no evidence of accumulation in the nasal turbinates, a site of oncogenicity for alachlor in the rat (Heydens et al. 2000). The ARP web site claims that both the ESA and OA degradates exhibit a low degree of toxicity to mammals and humans. The ESA degradates of chloroacetamides have, nonetheless, been shown to undergo microbial transformation to hydroxy-substituted acetamides (Laue, Field and Cook 1996); such degradates may thus pose an indirect risk.

Because of their likely widespread occurrence and the possible risk to human health posed by chloro-*s*-triazine and chloroacetamide herbicides as well as their degradates, all such herbicides not already regulated *and* their degradates have been included in EPA's Contaminant Candidate List (CCL) of species that may be subject to future regulation in drinking water (EPA 1998b). The presence of chloroacetanilide degradates in drinking water, and the ability of drinking water treatment practices to remove them, represents a potentially important "emerging issue" that has received remarkably little attention in the past.

REMOVAL OF HERBICIDES DURING DRINKING WATER TREATMENT

Conventional treatment processes, such as coagulation, chlorination, oxidation with permanganate, filtration, and aeration are known to be ineffective for removing chloro-*s*-triazine and chloroacetamide herbicides in drinking water, as summarized in a recent EPA review (EPA 2001a). Results of a prior study on this topic, in which pesticide removal was examined at the bench scale or in three full-scale water treatment plants, are summarized in Table 1.5. Coagulation with

Table 1.5
Summary of herbicide removal via conventional processes. Negative values can be attributed to analytical precision or to the inability to truly sample on a plug-flow basis.

	Removal efficiency (%)				
Treatment method	Atrazine	Simazine	Cyanazine	Alachlor	Metolachlor
Coagulation (in jar tests)	0*	-4*		4†	11‡
Softening-clarification (full-scale treatment plants)	-4	-10	-4	-5	-3
Chlorination (full-scale treatment plants)	≤6	≤7	≤3	≤9	≤3

Source: Data from Miltner et al. 1989
* 20 mg/L alum
† 15 mg/L alum
‡ 30 mg/L alum

alum only removes 4% of alachlor, 11% of metolachlor, and does not effect any removal of atrazine or simazine in jar test studies of surface water (Miltner et al. 1989). Softening and clarification does not produce measurable removal of these four pesticides (Miltner et al. 1989), no doubt stemming from their slow rates of basic hydrolysis. Chlorination was also shown (Miltner et al. 1989) to be ineffective at removing chloroacetamide and chloro-*s*-triazine herbicides at full-scale treatment plants.

A second study (EPA 2001a) investigated the removal of alachlor by chemical oxidation in laboratory settings; results are summarized in Table 1.6. Of various disinfectants tested, only ozone was found to remove alachlor, with removal efficiencies ranging from 75-97% in distilled water, groundwater and surface water; Cl_2, ClO_2, H_2O_2, and $KMnO_4$ were largely ineffective in removing this herbicide. $KMnO_4$ is also ineffective in atrazine removal (Randtke et al. 1994). The reaction of alachlor with ozone has been the subject of several prior investigations (Somich et al. 1988; Yao and Haag 1991; Beltrán, Acedo and Rivas 1999a; Beltrán, Acedo and Rivas 1999b; Beltrán et al. 2000). Ozonation (at a dosage of 2.2 mg/L) has been shown to effect removal of 83% of metolachlor in Dutch surface water (Kruithof et al. 1994). Although we have been unable to locate studies of acetochlor reaction with ozone, it is safe to assume removal efficiencies comparable to alachlor or metolachlor.

Ozonation of chloro-*s*-triazine herbicides has been the subject of somewhat more intensive investigations (Yao and Haag 1991; Adams and Randtke 1992; Hapeman-Somich et al. 1992; Koga, Kadokami and Shinohara 1992; Kruithof et al. 1994; Beltrán, Rivas and Acedo 1999; Acero, Stemmler and von Gunten 2000). O_3 reacts relatively slowly with atrazine; efficient treatment via ozonation hence typically requires combination with UV or hydrogen peroxide (*i.e.*, advanced oxidation processes; Acero, Stemmler and von Gunten 2000; Nelieu, Kerhonas and Einhorn 2000) that are not routinely employed for drinking water treatment in the U. S. Ozonation of atrazine results in a rather complex distribution of products, including 2-chloro-4-ethylimino-6-isopropylamino-*s*-triazine (the derivative of atrazine in which the ethylamino group $-NHCH_2CH_3$ has been converted to an ethylimino group $-N{=}CHCH_3$), 4-acetamido-2-chloro-6-isopropylamino-*s*-triazine (in which $-NHCH_2CH_3$ has undergone oxidation to $-NHC({=}O)CH_3$),

Table 1.6
Removal of alachlor by chemical oxidation during laboratory tests

Oxidant	Oxidant dose (mg/L)	Alachlor concentration (μg/L) and matrix	Contact time (h)	% Removal
Ozone	6.9	139 (distilled water)	0.22	95
	2.6-9.3	145 (groundwater)	0.22	79-96
	2.3-13.7	0.39-5.0 (surface water)	0.22	75-97
Chlorine	4.0-6.0	31-61 (surface water)	2.5-5.83	0-5
ClO_2	10.0	58 (distilled water)	22.3	0
H_2O_2	10.0	58 (distilled water)	22.3	0
$KMnO_4$	10.0	58 (distilled water)	22.3	0

Source: Data from EPA 2001a

deisopropylatrazine, and deethylatrazine, in decreasing order of abundance (Acero, Stemmler and von Gunten 2000). Because of health risks associated with these chloro-*s*-triazine degradates (EPA 2002a), ozonation should not be viewed as eliminating the health risks associated with triazine herbicides in drinking water.

One of the more successful herbicide treatment processes is adsorption onto carbon. Granular activated carbon (GAC) may be employed as a filter adsorber for taste and odor control, while post-filter adsorbers may be employed for removal of synthetic organic contaminants. Powdered activated carbon (PAC) is generally added within conventional treatment systems before or during the coagulation/flocculation and sedimentation steps. Results obtained using PAC during full-scale treatment indicate removal efficiencies for atrazine ranging from 28-87%, and 36-94% for alachlor, with better removal generally associated with higher PAC doses (Miltner et al. 1989; EPA 2001a). GAC removal efficiencies fall within the range of values obtained with PAC (Miltner et al. 1989; EPA 2001a). Although these results suggest that adsorption onto activated carbon can prove useful in reducing herbicide concentrations (in the case of PAC, at sufficiently high dosages), this treatment technology is rarely used by smaller water supply systems.

REMOVAL OF HERBICIDE DEGRADATES DURING DRINKING WATER TREATMENT

Some information exists pertaining to chloro-*s*-triazine degradate removal during water treatment. For example, isotherm studies reveal that although dealkylated chloro-*s*-triazine degradates can be treated using PAC, they adsorb much less readily than the parent compounds, leading to higher estimated costs (Adams and Watson 1996). Much less is known about the effectiveness of conventional treatment processes for removal of chloroacetamide degradates, although it is possible to develop "educated guesses" based on our understanding of the governing physical and chemical processes, coupled with information pertaining to chemical behavior derived from structural considerations. Coagulation and flocculation is most likely to remove hydrophobic contaminants, contaminants of relatively low molecular weight possessing acidic functional groups (*e.g.*, carboxylic acids), or high molecular weight compounds. At least some removal of chloroacetamide OA derivatives might be expected during this treatment step. Little or no removal of the majority of neutral chloroacetamide degradates, for which log K_{ow} values are typically less than for the parent compounds (Tables 1.1 and 1.4), would be expected. Acidification of water samples prior to coagulant addition might promote conversion of some herbicide degradates (especially the primary anilines) to cationic forms, which may prove to be more readily removed during coagulation/flocculation.

Reaction of certain chloroacetamide degradates with oxidants is likely to prove more rapid than is the case with the parent herbicides. For example, primary anilines (0.1 mM) react to completion within 10 minutes with 0.14 mM aqueous chlorine (Hwang, Larson and Snoeyink 1990), and within 2 minutes with 0.2 mM aqueous ozone (Chan and Larson 1991). Unfortunately, toxic or highly mutagenic disinfection byproducts are generated in the process. For example, primary anilines react with aqueous chlorine in homogeneous solution to form monochlorinated anilines (as the major products); in the presence of GAC, the major chlorination products shift to azobenzenes (Hwang, Larson and Snoeyink 1990). Acetylated derivatives (ring-substituted acetanilides) also form (Hwang, Larson and Snoeyink 1990). Reaction of aniline with aqueous ozone has been found to generate nitrosobenzene, nitrobenzene, azobenzene, and *N*-substituted polyaromatic compounds (Chan and Larson 1991), many of which are potent mutagens. This suggests removal of primary anilines would be highly desirable prior to disinfection.

Prior studies reveal amides to be relatively unreactive with chlorine (Katz 1986); chloroacetamide degradates retaining the acetamide functional group are hence unlikely to react with chlorine, though they may still react with aqueous ozone. Secondary anilines retaining the *N*-alkoxylalkyl group, but in which the amide bond has been cleaved, may represent an intermediate case.

As with the parent compounds, sorption onto GAC or PAC may represent the most promising treatment technology for chloroacetamide degradates in drinking water. Although removal efficiencies are likely in most cases to be lower than for the more hydrophobic parents, it may still be possible to achieve satisfactory removal. Recently, laboratory studies of ESA and OA degradate removal via PAC were conducted. Using Nashville, IL source water, 90% of parent chloroacetamides, 44% of ESA degradates, and 39% of OA degradates could be removed with a contact time of 60 minutes and a PAC dose of 20 mg/L (Gustafson et al. 2003). Although natural organic matter reduced the PAC's adsorption capacity, the target analytes did not appear to compete with one another for adsorption sites (Gustafson et al. 2003).

SPECIFIC RESEARCH OBJECTIVES

The preceding background section indicates important data gaps that pertain to the potential human health risk posed by herbicide degradates in drinking water. The most glaring areas of ignorance are associated with the occurrence of neutral chloroacetamide degradates, as well as techniques that might prove useful in their removal during water treatment. Existing evidence pertaining to toxicity and likely occurrence suggests many such could be present at concentrations sufficient to be of human health concern. We propose to redress these deficiencies by examining the occurrence of chloroacetamide degradates in raw and finished drinking water at a variety of Midwestern U.S. plants using conventional treatment trains. In order to provide a more complete understanding of the issue, our studies will also include analysis of the parent chloroacetamides, as well as more limited analyses of the ESA and OA degradates that are already recognized as common degradates (albeit of lesser health concern). The results will be used to develop hypotheses concerning specific treatment processes that may prove useful in treating chloroacetamide degradates in drinking water. These will be tested at the bench scale. Finally, even though the primary focus of this work is on the neutral chloroacetamide degradates of alachlor, acetochlor, and metolachlor, we will also extend our analyses of raw and finished drinking water samples to include the common chloro-*s*-triazine herbicides (atrazine, cyanazine, and simazine), as well as deethylatrazine, and deisopropylatrazine, (these two also originating from simazine). This will facilitate development of a fuller understanding of human health risks associated with the presence of herbicides and their degradates in drinking water.

This research will provide answers to the following questions:

- How can important chloroacetamide herbicide degradates be quantified in drinking water using simple but robust and widely available GC/MS techniques?
- What are the concentrations of common chloro-*s*-triazine and chloroacetamide herbicide degradates (relative to concentrations of the parent compounds) in raw and finished drinking water?
- How do differences in removal efficiency for chloroacetamide herbicide degradates relate to differences in treatment processes employed by various utilities?

These studies will provide regulators and utilities with information key to assessing whether current treatment practices suffice to eliminate health hazards that may be posed by specific chloroacetamide degradates (potentially leading to their elimination from future CCLs). Alternatively, if later studies confirm that such degradates pose a risk to human health and the present studies also indicate their widespread occurrence in raw water supplies, the efforts proposed herein will provide the drinking water community with information pertaining to specific technologies (such as, for example, a change in flocculant or disinfectant or pH) that could improve removal efficiencies. They may indicate a need for more advanced treatment processes (such as involving membranes or advanced oxidation). This work should thus provide some of the key pieces of information needed to resolve the complex question of whether herbicide degradates are of concern in drinking water.

CHAPTER 2
MATERIALS AND METHODS

REFERENCE STANDARDS

Chemicals

Alachlor (**I**) [15972-60-8; 2-chloro-2',6'-diethyl-*N*-(methoxymethyl)acetanilide], metolachlor (**XI**) [51218-45-2; 2-chloro-2'-ethyl-6'-methyl-*N*-(2-methoxy-1-methylethyl)acetanilide], acetochlor (**XX**) [34256-82-1; 2-chloro-2'-ethyl-6'-methyl-*N*-(ethoxymethyl)acetanilide], dimethenamid (**XXIX**) [87674-68-8; 2-chloro-*N*-(2,4-dimethyl-3-thienyl)-*N*-(2-methoxy-1-methylethyl)acetamide], atrazine (**XXXI**) [1912-24-9; 2-chloro-4-ethylamino-6-isopropylamino-*s*-triazine], desethyl atrazine (**XXXII**) [6190-65-4; 2-chloro-4-amino-6-isopropylamino-*s*-triazine], desisopropyl atrazine (**XXXIII**) [1007-28-9; 2-chloro-4-ethylamino-6-amino-*s*-triazine], simazine (**XXXIV**) [122-34-9; 2-chloro-4,6-diethylamino-*s*-triazine], cyanazine (**XXXV**) [21725-46-2; 2-chloro-4-ethylamino-6-methylpropionitrileamino-*s*-triazine], and deschloroacetylmetolachlor propanol (**XVII**) [2-[(2-ethyl-6-methylphenyl)amino]-1-propanol] were obtained from Chem Service (West Chester, PA). 2,6-Diethylaniline (**VIII**) [579-66-8] and 2-ethyl-6-methylaniline (**XXVIII**) [24549-06-2] were obtained from Aldrich. Alachlor OA (**IX**) [2-[(2,6-diethylphenyl)(methoxymethyl) amino]-2-oxoacetic acid], alachlor ESA (**X**) [2-[(2,6-diethylphenyl)(methoxymethyl) amino]-2-oxoethanesulfonic acid], and acetochlor OA (**XXIII**) [2-[(2-ethyl-6-methylphenyl)(ethoxymethyl)amino]-2-oxoacetic acid] were donated by Monsanto (St. Louis, MO). Metolachlor OA (**XVIII**) [2-[(2-ethyl-6-methylphenyl)(2-methoxy-1-methylethyl)amino]-2-oxoacetic acid] and metolachlor ESA (**XIX**) [2-[(2-ethyl-6-methylphenyl)(2-methoxy-1-methylethyl)amino]-2-oxoethanesulfonic acid] were donated by Syngenta (Greensboro, NC). Acetochlor ESA (**XXIV**) [2-[(2-ethyl-6-methylphenyl)(ethoxymethyl)amino]-2-oxoethanesulfonic acid] was obtained from the EPA National Standard Pesticide Repository (Ft. Meade, MD).

The following compounds were synthesized in our laboratory: hydroxyalachlor (**II**) [2-hydroxy-2',6'-diethyl-*N*-(methoxymethyl)acetanilide], deschloroalachlor (**III**) [2',6'-diethyl-*N*-(methoxymethyl)acetanilide], 2-chloro-2'-6'-diethylacetanilide (**IV**), 2-hydroxy-2'-6'-diethylacetanilide (**V**), 2-hydroxy-2'-6'-diethyl-*N*-methylacetanilide (**VI**), 2'-6'-diethylacetanilide (**VII**) [16665-89-7], hydroxymetolachlor (**XII**) [2-hydroxy-2'-ethyl-6'-methyl-*N*-(2-methoxy-1-methylethyl)acetanilide], deschlorometolachlor (**XIII**) [2'-ethyl-6'-methyl-*N*-(2-methoxy-1-methylethyl)acetanilide], a morpholinone derivative of metolachlor (**XIV**) [4-(2-ethyl-6-methylphenyl)-5-methyl-3-morpholinone], metolachlor propanol (**XV**) [2-chloro-2'-ethyl-6'-methyl-*N*-(2-hydroxy-1-methyl-ethyl)acetanilide], deschloroacetylmetolachlor (**XVI**) [2'-ethyl-6'-methyl-*N*-(2-methoxy-1-methylethyl)aniline], hydroxyacetochlor (**XXI**) [2-hydroxy-2'-ethyl-6'-methyl-*N*-(ethoxymethyl)acetanilide], deschloroacetochlor (**XXII**) [2'-ethyl-6'-methyl-*N*-(ethoxymethyl)acetanilide], 2-chloro-2'-ethyl-6'-methylacetanilide (**XXV**), 2-hydroxy-2'-ethyl-6'-methylacetanilide (**XXVI**), 2'-ethyl-6'-methylacetanilide (**XXVII**), and deschlorodimethenamid (**XXX**) [*N*-(2,4-dimethyl-3-thienyl)-*N*-(2-methoxy-1-methylethyl)acetamide]. Synthesis procedures and both ^{1}H NMR (proton nuclear magnetic resonance) spectral data and electron ionization mass spectra data are provided below.

Surrogate standards for the neutral compounds were ring labeled $^{13}C_6$-metolachlor and $^{13}C_3$-atrazine, obtained from Cambridge Isotope Labs (Andover, MA); the surrogate for the ionic

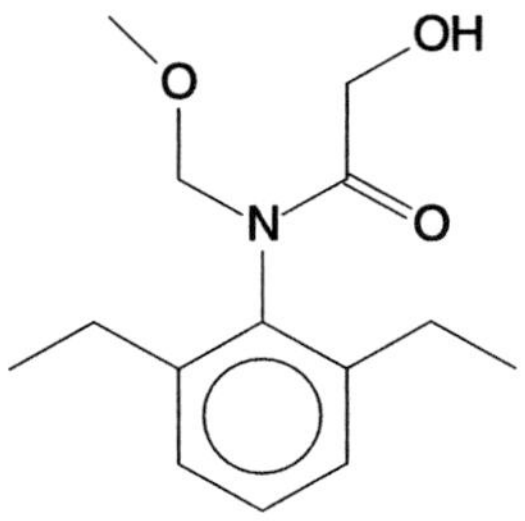

Figure 2.1 Structure of hydroxyalachlor (II)

compounds was 2-benzoylbenzoic acid, obtained from Aldrich (Milwaukee, WI). Internal standards for compounds analyzed via GC/MS were acenaphthene-d_{10}, anthracene-d_{10}, and chrysene-d_{12}, obtained from Cambridge Isotope Labs. The internal standard for the ESA degradates, which were analyzed via high pressure liquid chromatography with diode-array detection (HPLC-DAD), was 2,4-dichlorophenylacetic acid (Aldrich).

Synthesis Procedures

The following briefly describes methods used to synthesize the chloroacetamide degradates sought in the following chapters. All ^{1}H NMR (proton nuclear magnetic resonance) spectra were obtained using a Varian Unity 400 MHz FT-NMR after dissolving compounds in $CDCl_3$ containing tetramethylsilane (TMS). Concentrations were ~500 mg/L for NMR samples. Chemical shifts are reported as ppm referenced to TMS. All mass spectra were obtained using a ThermoQuest Trace 2000 gas chromatograph with a programmed temperature vaporization injector (PTV) coupled to a quadrupole mass spectrometer (GC/MS) in electron ionization mode. Concentrations were ~500 mg/L for NMR samples. Further details of the synthesis procedures and characterization can be found free of charge via the internet as part of Hladik, Hsiao and Roberts (2005).

II. Hydroxyalachlor [2-hydroxy-2',6'-diethyl-N-(methoxymethyl)acetanilide]

Alachlor (as much as would dissolve in the aqueous solution, no co-solvent used) was allowed to sit at room temperature in 2 N aqueous NaOH in the dark until the reaction was complete (several months). Upon completion of the reaction, the product was extracted with dichloromethane (50 mL) and dried under nitrogen. The product is a colorless oily liquid with purity >95% (determined by GC/MS). Structure is shown in Figure 2.1. MS *m/z* (relative intensity): 251 $[M]^{+\cdot}$ (0), 219 $[M\text{-}OCH_3\text{-}H]^+$ (19.7), 188 $[M\text{-}OCH_3, CH_2OH\text{-}H]^{+\cdot}$ (100), 160 $[M\text{-}OCH_3, COCH_2OH\text{-}H]^{+\cdot}$ (94.3). ^{1}H NMR ($CDCl_3$) δ 7.34 (t, 1H, CHC*H*CH), 7.18 (d, 2H, C*H*CHC*H*), 4.95 (d, 2H, CH_2OCH_3), 3.61 (d, 2H, $COCH_2OH$), 3.49 (s, 3H, CH_2OCH_3), 3.28 (t, 1H, OH), 2.50 (m, 4H, 2-CH_3CH_2), 1.17 (t, 6H, 2-CH_3CH_2).

III. Deschloroalachlor [2',6'-diethyl-N-(methoxymethyl)acetanilide]

Procedure was based on one reported by Eykholt and Davenport (1998). To a solution of 1:1 water:acetone (100 mL) containing 2 g Fe(0) powder 100 mg of alachlor was added. The

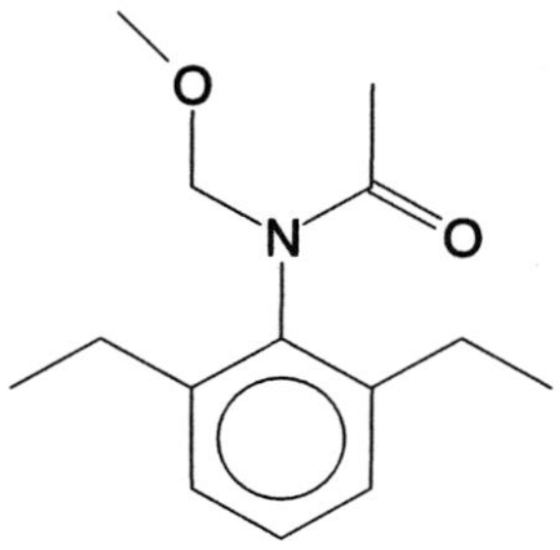

Figure 2.2 Structure of deschloroalachlor (III)

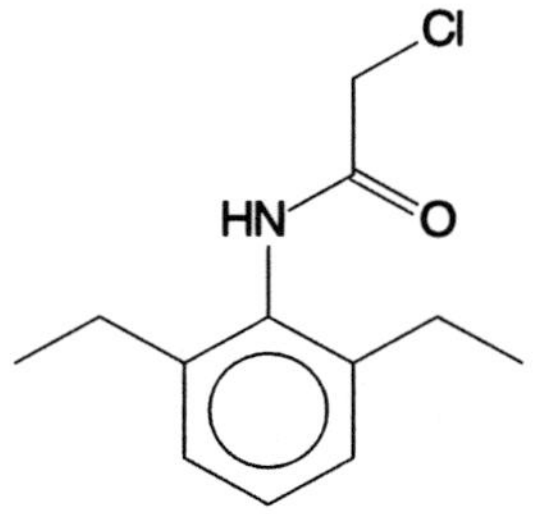

Figure 2.3 Structure of 2-chloro-2'-6'-diethylacetanilide (IV)

reductive dehalogenation occurred via a surface mediated reaction that results in replacement of chlorine by hydrogen. Once the reaction was complete, the product was extracted with hexane (50 mL). The final product (on evaporation of the hexane) was a white powder with >95% purity (determined by GC/MS). Structure is shown in Figure 2.2. MS *m/z* (relative intensity): 235 $[M]^{+\cdot}$ (26.7), 203 $[M\text{-}OCH_3\text{-}H]^+$ (42.4), 178 $[M\text{-}COCH_3, CH_3\text{+}H]^{+\cdot}$ (60.4), 161 $[M\text{-}COCH_3, OCH_3]^+$ (100). ^{1}H NMR ($CDCl_3$) δ 7.26 (t, 1H, CHC*H*CH), 7.17 (d, 2H, C*H*CHC*H*), 4.89 (s, 2H, CH_2OCH_3), 3.43 (s, 3H, CH_2OCH_3), 2.54 (m, 4H-, 2-CH_3CH_2), 1.72 (s, 3H, $COCH_3$), 1.22 (t, 6H, 2-CH_3CH_2).

IV. 2-Chloro-2'-6'-diethylacetanilide

Procedure was based on that given by Potter and Carpenter (1995). To a solution of 1:1 acetone/ 3 N aqueous HCl (250 mL), 200 mg of alachlor was added and refluxed for 24 hours. This effects the loss of the *N*-alkoxymethyl group. Once the product was formed (after several hours), the solution was cooled, NaCl was added to saturate the solution, and the product was extracted with dichloromethane (100 mL). The final product (obtained on evaporation of the dichloromethane) was a colorless crystal with >95% purity (determined by GC/MS). Structure is shown in Figure 2.3. MS *m/z* (relative intensity): 225 $[M]^{+\cdot}$ (13.3), 176 $[M\text{-}CH_2Cl]^{+\cdot}$ (100), 148 $[M\text{-}COCH_2Cl]^{+\cdot}$ (19.6). ^{1}H NMR ($CDCl_3$) δ 7.85 (s, 1H, NH), 7.28-7.10 (m, 3H, phenyl), 4.25 (s, 2H, $COCH_2Cl$), 2.60 (q, 4H, 2-CH_3CH_2), 1.22 (t, 6H, 2-CH_3CH_2).

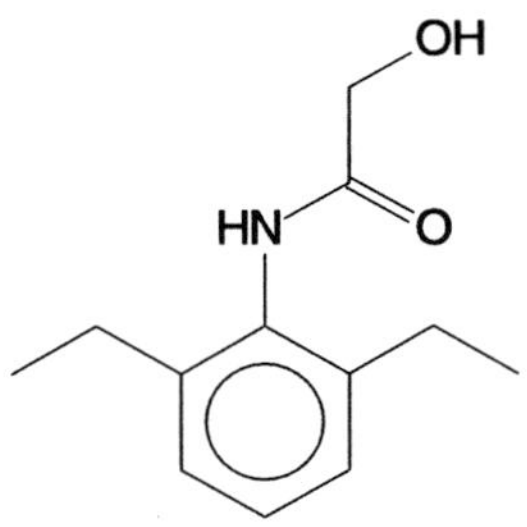

Figure 2.4 Structure of 2-hydroxy-2'-6'-diethylacetanilide (V)

V. 2-Hydroxy-2'-6'-diethylacetanilide

Procedure was based upon that given by Potter and Carpenter (1995). A solution of 35:65 acetone/ 0.7 N aqueous NaOH (100 mL), containing 100 mg of 2-chloro-2'-6'-diethylacetanilide, was heated at reflux for 8 hours. The solution was cooled, saturated with NaCl, and the product was then extracted with dichloromethane (50 mL). The white powder (obtained on evaporation of the dichloromethane) had a final purity of >95% (determined by GC/MS). Structure is shown in Figure 2.4. MS *m/z* (relative intensity): 207 $[M]^{+\cdot}$ (7), 176 $[M\text{-}CH_2OH]^{+\cdot}$ (100), 148 $[M\text{-}COCH_2OH]^{+\cdot}$ (15.1). ^1H NMR ($CDCl_3$) δ 7.85 (s, 1H, NH), 7.25-7.10 (m, 3H, phenyl), 4.35 (d, 2H, $COCH_2OH$), 3.48 (s, 1H, OH), 2.60 (q, 4H, $2\text{-}CH_3CH_2$), 1.22 (t, 6H, $2\text{-}CH_3CH_2$).

VI. 2-Hydroxy-2'-6'-diethyl-N-methylacetanilide

Procedure was based on that of Chiron et al. (1995). In 50 mL of dichloromethane (DCM) with 1.4 mL of triethylamine and 0.55 mL 2,6-diethylaniline, 0.67 mL acetoxyacetylchloride was added. The reaction was allowed to proceed at room temperature for 21 hours. The reaction mixture was then washed with 25 mL water and dried over $MgSO_4$. The remaining DCM solution was evaporated. To 0.10 g of the above product in 20 mL anhydrous tetrahydrofuran (THF), 0.15 g of NaH in 3 mL anhydrous THF was added. The reaction was allowed to proceed for 1 hour at room temperature before 0.20 mL of CH_3I in 10 mL of THF was added, at which point the reaction was allowed to continue for an additional 2 hours. The remaining solution was blown dry. To this product in 12 mL of THF, 25 mL of 1% aqueous NaOH was added and reacted for 15 min. Work-up yielded the product, a brownish oily liquid at >90% purity (as determined by GC/MS). Structure is shown in Figure 2.5. MS *m/z* (relative intensity): 221 $[M]^{+\cdot}$ (2.9), 190 $[M\text{-}CH_2OH]^{+\cdot}$ (100), 162 $[M\text{-}COCH_2OH]^{+\cdot}$ (19.4). ^1H NMR ($CDCl_3$) δ 7.38-7.10 (m, 3H, phenyl), 3.59 (d, 2H, $COCH_2OH$), 3.39 (s, 1H, OH), 3.24 (s, 3H, NCH_3), 2.51 (m, 4H, $2\text{-}CH_3CH_2$), 1.24 (t, 6H, $2\text{-}CH_3CH_2$).

VII. 2'-6'-Diethylacetanilide

Procedure was based on that given by Nesnow et al. (1995). In anhydrous ethyl acetate (125 mL) under nitrogen 17 g of 2,6-diethylaniline was dissolved. To this mixture 25 mL of acetylchloride was added dropwise and the solution was heated at reflux for 2 hours under nitrogen. After the reaction was complete the ethyl acetate was distilled off and 100 mL ice-cold

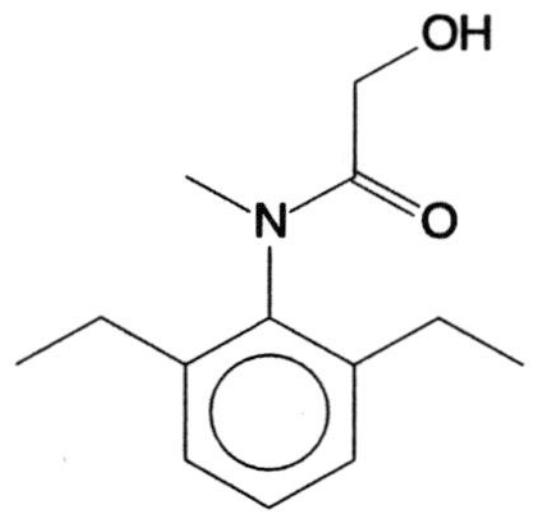

Figure 2.5 Structure of 2-hydroxy-2'-6'-diethyl-*N*-methylacetanilide (VI)

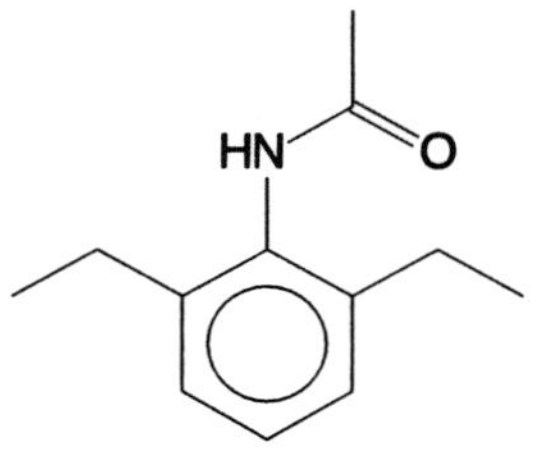

Figure 2.6 Structure of 2'-6'-diethylacetanilide (VII)

water was added to the oily residue. This solution was extracted with dichloromethane (150 mL). The organic layer was washed with water (3 × 125 mL), dried over $MgSO_4$, filtered and then concentrated by evaporation. The final product was isolated by extracting the resulting solid with dichloromethane (50 mL). The product (obtained on evaporation of the dichloromethane) was a white powder with a purity >95% (determined by GC/MS). Structure is shown in Figure 2.6. MS *m/z* (relative intensity): 191 $[M]^{+\cdot}$ (54.1), 148 $[M\text{-}COCH_3]^{+\cdot}$ (100), 134 $[M\text{-}COCH_3, N]^+$ (91.9). ^1H NMR ($CDCl_3$) δ 7.25-7.09 (m, 3H, phenyl), 6.65 (s, 1H, NH), 2.71 (dq, 4H, 2-CH_3CH_2), 2.22 (s, 3H, $COCH_3$), 1.20 (t, 6H, 2-CH_3CH_2).

XII. Hydroxymetolachlor [2-hydroxy-2'-ethyl-6'-methyl-N-(2-methoxy-1-methyl-ethyl)acetanilide]

To 30 mL of glycol dimethyl ether, 1.6 g of sodium acetate and 0.1 g NaI were added. To this solution 200 mg of metolachlor was added. The solution was refluxed at 95 °C with stirring for 5-7 hours. The liquid was decanted and the solvent evaporated. A 5-mL aliquot of 1 N aqueous NaOH was added immediately and the reaction was left overnight. The final product, a yellowish oily residue, was extracted with dichloromethane. Final purity >95% (determined by GC/MS). Structure is shown in Figure 2.7. MS *m/z* (relative intensity): 265 $[M]^{+\cdot}$ (0), 220 $[M\text{-}CH_2OCH_3]^{+\cdot}$ (40.7), 193 $[M\text{-}CH(CH_3)CH_2OCH_3\text{+}H]^+$ (13.1), 162 $[M\text{-}CH(CH_3)CH_2OCH_3, CH_2OH\text{+}H]^{+\cdot}$ (100). ^1H NMR ($CDCl_3$) δ 7.25-7.07 (m, 3H, phenyl), 4.35 (m, 1H, $CH(CH_3)CH_2$), 3.80 (m, 2H, $CH(CH_3)CH_2$), 3.57 (d, 2H, $COCH_2OH$), 3.41(s, H, OH), 3.27 (s, 3H, OCH_3), 2.55 (m, 2H, CH_3CH_2), 2.23 (s, 3H, CH_3), 1.25 (t, 3H, CH_3CH_2), 1.10 (d, 3H, $CH(CH_3)CH_2$).

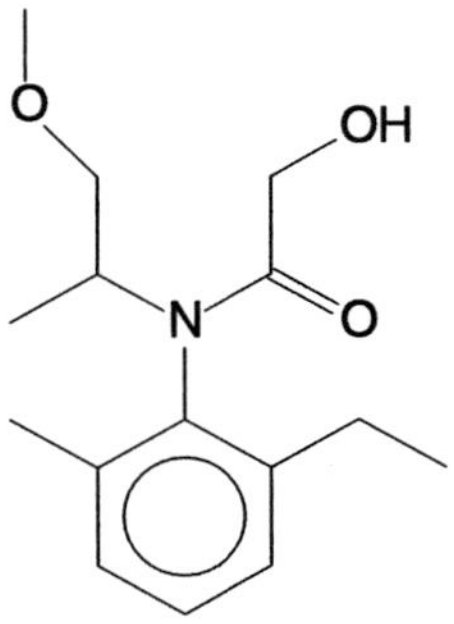

Figure 2.7 Structure of hydroxymetolachlor (XII)

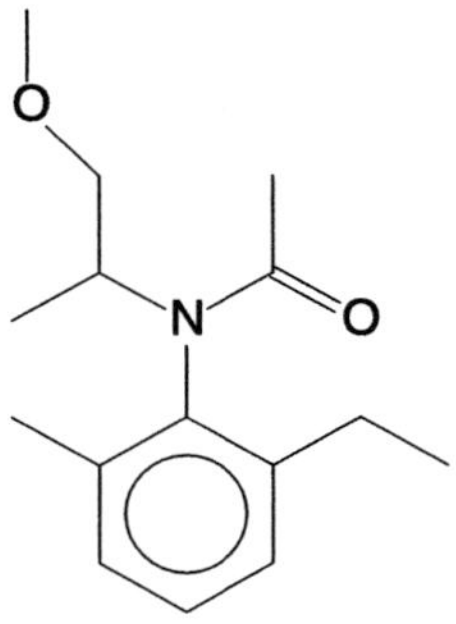

Figure 2.8 Structure of deschlorometolachlor (XIII)

XIII. Deschlorometolachlor [2'-ethyl-6'-methyl-N-(2-methoxy-1-methylethyl)acetanilide]

Procedure was based on one reported by Eykholt and Davenport (1998). To a solution of 1:1 water:methanol (100 mL) containing 2 g Fe(0) powder, 100 mg of metolachlor was dissolved. The reductive dehalogenation occurred via a surface-mediated reaction that results in replacement of chlorine by hydrogen. Once the reaction was complete, the product was extracted with hexane (50 mL). The final product (on evaporation of the hexane) was a white powder with >95% purity (determined by GC/MS). Structure is shown in Figure 2.8. MS *m/z* (relative intensity): 249 $[M]^{+\cdot}$ (0), 204 $[M\text{-}CH_2OCH_3]^{+\cdot}$ (28.1), 177 $[M\text{-}CH(CH_3)CH_2OCH_3\text{+}H]^{+}$ (7.2), 162 $[M\text{-}CH(CH_3)CH_2OCH_3, CH_3\text{+}H]^{+\cdot}$ (100), 133 $[M\text{-}CH(CH_3)CH_2OCH_3, COCH_3]^{+}$ (6.3). ^{1}H NMR ($CDCl_3$) δ 7.21-7.10 (m, 3H, phenyl), 4.24 (m, 1H, $CH(CH_3)CH_2$), 3.72-3.42 (2m, 2H, $CH(CH_3)CH_2$), 3.21 (2s, 3H, CH_2OCH_3), 2.55 (m, 2H, CH_3CH_2), 2.23 (2s, 3H, CH_3), 1.67 (2s, 3H, $COCH_3$), 1.21 (t, 3H, CH_3CH_2), 1.10 (2d, 3H, $CH(CH_3)CH_2$).

XIV. Metolachlor morpholinone [4-(2-ethyl-6-methylphenyl)-5-methyl-3-morpholinone]

To a solution of 6 N HCl (100 mL), 56 mg of metolachlor (dissolved in 1 mL of methanol) was added and incubated at 85 °C for 120 min. After cooling and extracting any unreacted metolachlor with hexane (100 mL), the aqueous solution was basified with 350 mL of 2 N aqueous NaOH. The aqueous solution was extracted with 200 mL of dichloromethane. The solvent was

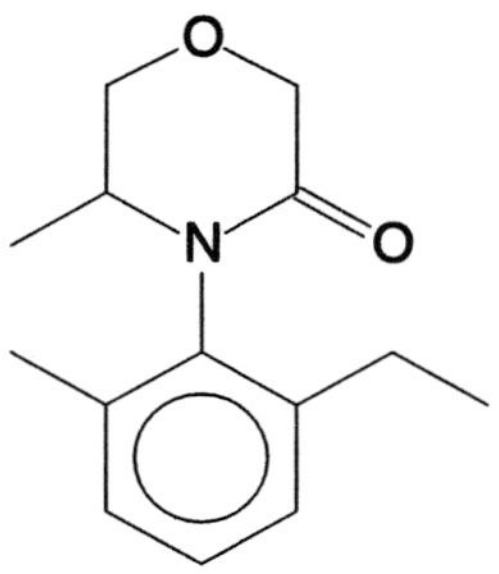

Figure 2.9 Structure of metolachlor morpholinone (XIV)

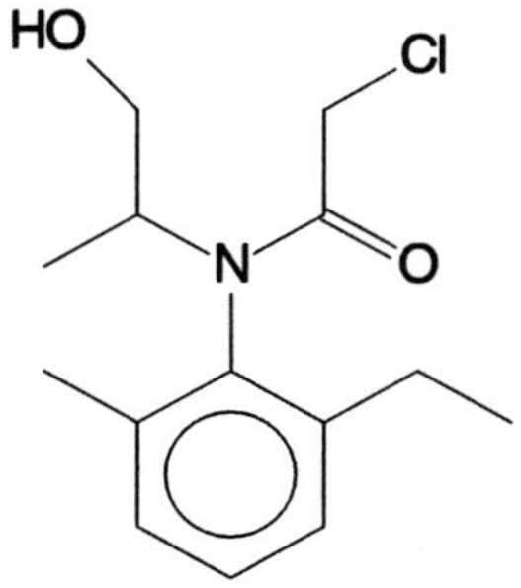

Figure 2.10 Structure of metolachlor propanol (XV)

blown down to give the product, a white powder, with >95% purity (determined by GC/MS). Structure is shown in Figure 2.9. MS *m/z* (relative intensity): 233 $[M]^{+\cdot}$ (50.3), 188 $[M\text{-}CH_2OCH_2\text{-}H]^{+\cdot}$ (48.8), 161 $[M\text{-}CH_2OCH_2CH(CH_3)]^{+}$ (83.0), 146 $[M\text{-}CH_2OCH_2CH(CH_3), O\text{+}H]^{+\cdot}$ (100). ^{1}H NMR ($CDCl_3$) δ 7.25-7.09 (m, 3H, phenyl), 4.39 (q, 2H, $NCOCH_2$), 4.05 (m, 1H, $NCH(CH_3)CH_2$), 3.75 (m, 2H, $NCH(CH_3)CH_2$), 2.57 (2q, 2H, CH_3CH_2), 2.25 (2s, 3H, CH_3), 1.26 (2t, 3H, CH_3CH_2), 1.13 (2d, 3H, $NCH(CH_3)CH_2$).

XV. Metolachlor propanol [2-chloro-2'-ethyl-6'-methyl-N-(2-hydroxy-1-methyl-ethyl)acetanilide]

To 100 mL of 6 N aqueous HCl, 54 mg of metolachlor (dissolved in 1 mL methanol) was added and incubated at 85 °C for 120 min. After cooling and extracting the remaining metolachlor with hexane (100 mL), the aqueous solution was neutralized with 51.5 g of $NaCO_3$. After neutralization the product was extracted from the aqueous solution with 50 mL dichloromethane. The solvent was evaporated to give the product, a white powder, with >83% purity (determined by GC/MS). Structure is are shown in Figure 2.10. MS *m/z* (relative intensity): 269 $[M]^{+\cdot}$ (1.4), 238 $[M\text{-}CH_2OH]^{+\cdot}$ (49.4), 162 $[M\text{-}CH_2OH, \text{-}COCH_2Cl \text{+}H]^{+\cdot}$ (100), 146 $[M\text{-}CH_2OH, CH_3, COCH_2Cl]^{+\cdot}$ (56.3). ^{1}H NMR ($CDCl_3$) δ 7.26-7.10 (m, 3H, phenyl), 4.05 (m, H, $CH(CH_3)CH_2$), 3.87 (m, 2H, $CH(CH_3)CH_2$), 3.66 (2q, 2H, $COCH_2Cl$), 3.45 (s, 1H, OH), 2.60 (2q, 2H, CH_3CH_2), 2.25 (2s, 3H, CH_3), 1.25 (2t, 3H, CH_3CH_2), 1.18 (2d, 3H, $CH(CH_3)CH_2$).

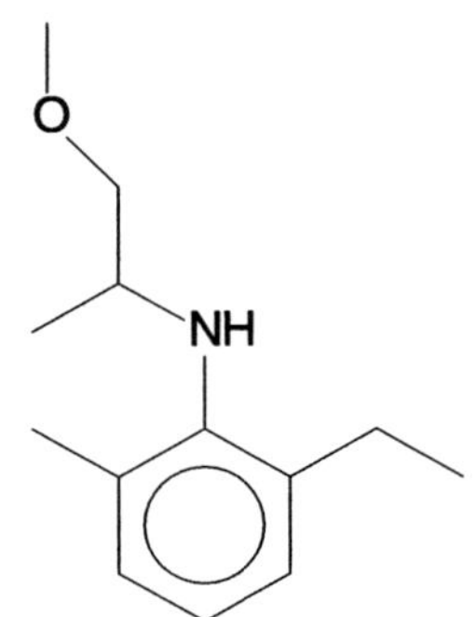

Figure 2.11 Structure of deschloroacetylmetolachlor (XVI)

XVI. Deschloroacetylmetolachlor [2'-ethyl-6'-methyl-N-(2-methoxy-1-methylethyl)aniline]

To a 127 mL of pyridine, 53 mg tosyl chloride was dissolved. To this mixture 28 mL of 1-methoxy-2-propanol was added dropwise while the solution was in an ice bath. The ice bath was removed and the solution was stirred overnight. After stirring the solution was extracted with 250 mL DI water. 200 mL of the aqueous solution was extracted with ether 3 times (30 mL: 30 mL: 15 mL). The ether extracts were combined and washed with 2 N aqueous H_2SO_4 until the wash came through acidic. The ether solution was dried with sodium sulfate and left to evaporate leaving 1-methoxy-2-tosylpropane. To 12 mL of the 1-methoxy-2-tosylpropane, 3 mL of 2-ethyl-6-methyl aniline was added, along with 2 mL of pyridine. The reaction was heated at 100 °C overnight to yield deschloroacetylmetolachlor. The product, a yellowish liquid, was further purified using a silica gel column. Final purity was >95% (determined by GC/MS). Structure is shown in Figure 2.11. MS *m/z* (relative intensity): 207 $[M]^{+\cdot}$ (6.8), 162 $[M\text{-}CH_2OCH_3]^{+\cdot}$ (100), 133 $[M\text{-}CH(CH_3)CH_2OCH_3\text{-}H]^+$ (17.5). ^1H NMR ($CDCl_3$) δ 7.03-6.83 (m, 3H, phenyl), 3.35 (m, 1H, $\mathit{CH}(CH_3)CH_2$), 3.34 (m, 2H, $\mathit{CH_2}OCH_3$), 3.32 (s, 3H, $CH_2O\mathit{CH_3}$), 2.60 (m, 2H, $CH_3\mathit{CH_2}$), 2.25 (s, 3H, $\mathit{CH_3}$), 1.20 (t, 3H, $\mathit{CH_3}CH_2$), 1.10 (m, 3H, $CH(\mathit{CH_3})CH_2$).

XXI. Hydroxyacetochlor [2-hydroxy-2'-ethyl-6'-methyl-N-(ethoxymethyl)acetanilide]

Procedure was based on that of Feng (1991). To 30 mL of glycol dimethyl ether, 1.6 g of sodium acetate and 0.1 g NaI were added. To this solution 200 mg of acetochlor was added. The solution was refluxed at 95 °C with stirring for 5-7 hours. The liquid was decanted and the solvent evaporated. A 5-mL aliquot of 1 N aqueous NaOH was added immediately and the reaction was left overnight. The final product, a yellow liquid, was extracted with dichloromethane (50 mL). Final purity was >95% (determined by GC/MS). Structure is shown in Figure 2.12. MS *m/z* (relative intensity): 251 $[M]^{+\cdot}$ (4.5), 205 $[M\text{-}OCH_2CH_3\text{-}H]^+$ (66.8), 174 $[M\text{-}OCH_2CH_3, CH_2OH\text{-}H]^{+\cdot}$ (100), 146 $[M\text{-}OCH_2CH_3, COCH_2OH\text{-}H]^{+\cdot}$ (98.6). ^1H NMR ($CDCl_3$) δ 7.25-7.12 (m, 3H, phenyl), 5.03 (dd, 2H, $\mathit{CH_2}OCH_2CH_3$), 4.26 (d, 2H, $CO\mathit{CH_2}OH$), 3.61 (s, H, OH), 3.50 (q, 2H, $CH_2O\mathit{CH_2}CH_3$), 2.54 (m, 2H, $\mathit{CH_3}CH_2$), 2.22 (s, 3H, CH_3), 1.23 (t, 3H, $\mathit{CH_3}CH_2$), 1.16 (q, 3H, $CH_2OCH_2\mathit{CH_3}$).

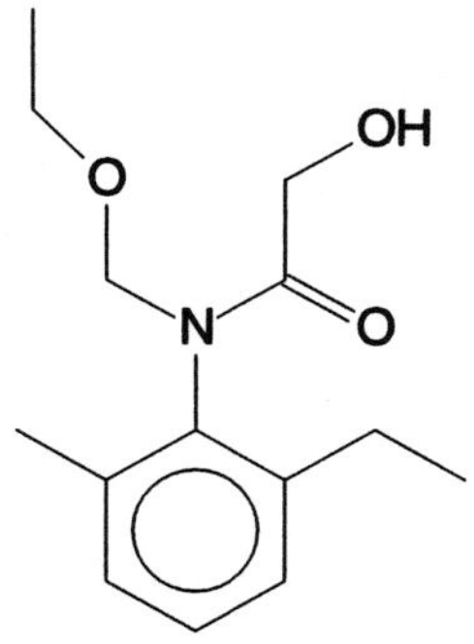

Figure 2.12 Structure of hydroxyacetochlor (XXI)

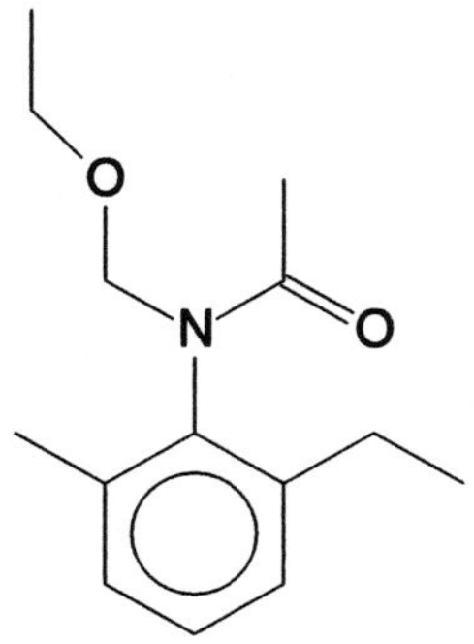

Figure 2.13 Structure of deschloroacetochlor (XXII)

XXII. Deschloroacetochlor [2'-ethyl-6'-methyl-N-(ethoxymethyl)acetanilide]

To a solution of 1:1 water:methanol (100 mL) containing 2 g Fe(0) powder, 100 mg of acetochlor was dissolved. The reductive dehalogenation occurred via a surface-mediated reaction that resulted in replacement of chlorine by hydrogen. Once the reaction was complete, the product was extracted with hexane (50 mL). The final product (obtained on evaporation of the hexane) was a white powder, >95% pure (determined by GC/MS). Structure is shown in Figure 2.13. MS *m/z* (relative intensity): 235 $[M]^{+\cdot}$ (10.8), 206 $[M\text{-}CH_2CH_3]^{+\cdot}$ (65.6), 189 $[M\text{-}CH_3, OCH_3]^+$ (26.6), 164 $[M\text{-}CH_2CH_3, COCH_3\text{+}H]^{+\cdot}$ (100). ^1H NMR ($CDCl_3$) δ 7.25-7.12 (m, 3H, phenyl), 4.98 (dd, 2H, $CH_2OCH_2CH_3$), 3.71 (q, 2H, $CH_2OCH_2CH_3$), 2.57 (m, 2H, CH_3CH_2), 2.16 (s, 3H, CH_3), 1.75 (s, 3H, $COCH_3$), 1.24 (t, 3H, CH_3CH_2), 1.18 (t, 3H, $CH_2OCH_2CH_3$).

XXV. 2-Chloro-2'-ethyl-6'-methylacetanilide

To 250 mL anhydrous ethyl acetate under nitrogen, 30 g of 2-ethyl-6-methyl aniline was added. To this mixture, 44 g of chloroacetylchloride was added dropwise and the solution was heated at reflux for 2 hours under nitrogen. After the reaction was complete, the ethyl acetate was distilled off and 200 mL ice-cold water was added to the oily residue. This solution was

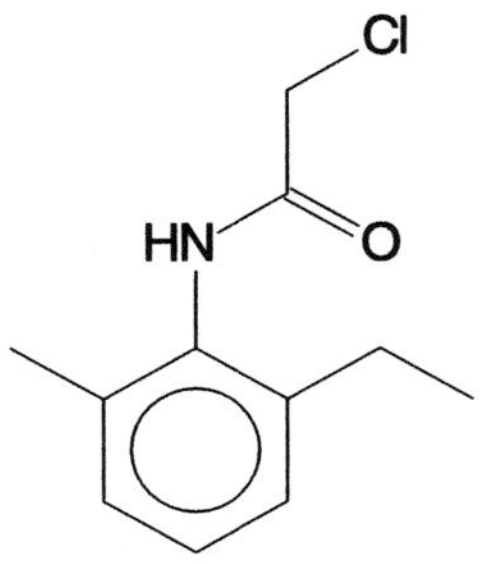

Figure 2.14 Structure of 2-chloro-2'-ethyl-6'-methylacetanilide (XXV)

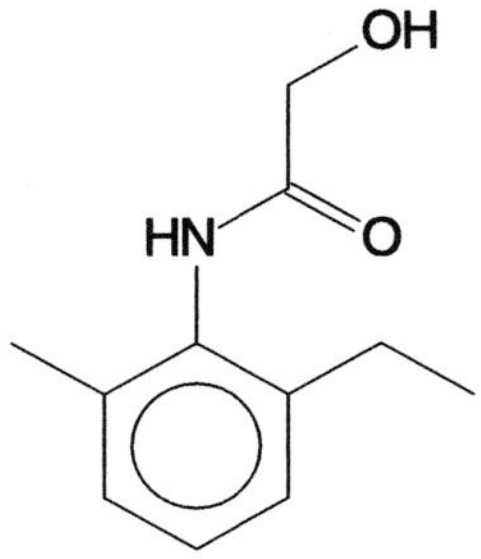

Figure 2.15 Structure of 2-hydroxy-2'-ethyl-6'-methylacetanilide (XXVI)

extracted with 300 mL of dichloromethane (DCM). The organic layer was washed with water (3 × 250 mL), dried over $MgSO_4$, filtered and then concentrated by evaporation. The desired product was isolated from its di-chloroacetylated side-product by adding 5 mL 2 N aqueous NaOH and conducting a subsequent DCM extraction (100 mL). The final product (obtained by evaporating the DCM) was a white powder, >95% pure (determined by GC/MS). Structure is shown in Figure 2.14. MS *m/z* (relative intensity): 211 $[M]^{+\cdot}$ (19.1), 162 $[M\text{-}CH_2Cl]^{+\cdot}$ (100), 134 $[M\text{-}COCH_2Cl]^{+\cdot}$ (26.5). ^1H NMR ($CDCl_3$) δ 7.85 (s, 1H, NH), 7.88-7.24 (m, 3H, phenyl), 4.25 (s, 2H, $COCH_2Cl$), 2.60 (q, 2H, CH_3CH_2), 2.20 (s, 3H, CH_3) 1.20 (t, 3H, CH_3CH_2).

XXVI. 2-Hydroxy-2'-ethyl-6'-methylacetanilide

To a solution of 35:65 acetone/ 0.7 N aqueous NaOH (100 ml), 100 mg of 2-chloro-2'-ethyl-6'-methylacetanilide was added and heated at reflux for 8 hours. The solution was cooled, saturated with NaCl, and the product was then extracted with dichloromethane (50 mL). The final product, a white powder, was >95% pure (determined by GC/MS). Structure is shown in Figure 2.15. MS *m/z* (relative intensity): 193 $[M]^{+\cdot}$ (14.5), 162 $[M\text{-}CH_2OH]^{+\cdot}$ (100), 134 $[M\text{-}COCH_2OH]^{+\cdot}$ (41.4). ^1H NMR ($CDCl_3$) δ 7.75 (s, 1H, NH), 7.23-7.04 (m, 3H, phenyl), 4.34 (d, 2H, $COCH_2OH$), 3.70 (s, 1H, OH), 2.60 (q, 2H, CH_3CH_2), 2.23 (s, 3H, CH_3), 1.15 (t, 3H, CH_3CH_2).

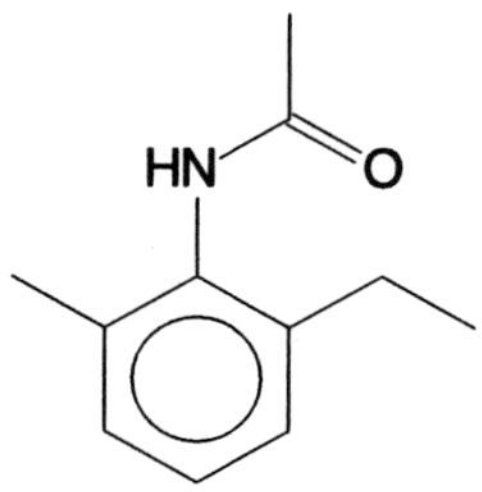

Figure 2.16 Structure of 2'-ethyl-6'-methylacetanilide (XXVII)

XXVII. 2'-Ethyl-6'-methylacetanilide

In 125 mL anhydrous ethyl acetate under nitrogen, 15 g of 2-ethyl-6-methyl aniline was added. To this mixture 25 mL of acetylchloride was added dropwise and the solution was heated at reflux for 2 hours under nitrogen. After the reaction was complete the ethyl acetate was distilled off and 100 mL ice-cold water was added to the oily residue. This solution was extracted with dichloromethane (150 mL). The organic layer was washed with water (3 × 125 mL), dried over $MgSO_4$, filtered and then concentrated by evaporation of the dichloromethane. The final product was isolated by extracting the resulting solid with dichloromethane. The product (obtained on evaporation of the dichloromethane) is a white powder with a purity >95% (determined by GC/MS). Structure is shown in Figure 2.16. MS *m/z* (relative intensity): 177 $[M]^{+\cdot}$ (38.4), 134 $[M\text{-}COCH_3]^{+\cdot}$ (77.0), 120 $[M\text{-}COCH_3, N]^+$ (100). ^1H NMR ($CDCl_3$) δ 7.25-7.06 (m, 3H, phenyl), 6.76 (s, 1H, NH), 2.62 (m, 2H, $CH_3\mathit{CH_2}$), 2.27 (m, 3H, CH_3), 2.23 (1.75) (2s, 3H, $COCH_3$), 1.10 (s, 3H, $\mathit{CH_3}CH_2$).

XXX. Deschlorodimethenamid [N-(2,4-dimethyl-3-thienyl)-N-(2-methoxy-1-methylethyl)acetamide]

To a solution of 1:1 water:methanol (100 mL) containing 2 g Fe(0) powder, 100 mg of dimethenamid was added. The reductive dehalogenation occurred via a surface-mediated reaction that results in replacement of chlorine by hydrogen. Once the reaction was complete, the product was extracted with hexane (50 mL). The final product (on evaporation of the hexane) was a white powder with >95% purity (determined by GC/MS). Structure is shown in Figure 2.17. MS *m/z* (relative intensity): 241 $[M]^{+\cdot}$ (1.9), 196 $[M\text{-}CH_2OCH_3]^{+\cdot}$ (58.2), 169 $[M\text{-}CH(CH_3)CH_2OCH_3 +H]^+$ (51.0), 154 $[M\text{-} CH(CH_3)CH_2OCH_3, CH_3 +H]^{+\cdot}$ (100). ^1H NMR ($CDCl_3$) δ 6.77 (s, 1H, SCHC), 4.68 (4.58) (m, 1H, $\mathit{CH}(CH_3)CH_2$), 3.54 (3.31) (m, 2H, $\mathit{CH_2}OCH_3$), 3.21 (3.23) (s, 3H, $CH_2O\mathit{CH_3}$), 2.06 (2.31) (d, 6H, 2-CH_3), 1.72 (s, 3H, $COCH_3$), 1.19 (1.06) (d, 3H, $CH(\mathit{CH_3})CH_2$).

HERBICIDE AND HERBICIDE DEGRADATE ANALYSIS METHOD FOR FIELD SAMPLES

Initial Recovery Studies

Initial recovery studies, the purpose of which was to determine the suitability of the SPE media employed, were conducted using deionized water as a sample matrix. In these studies,

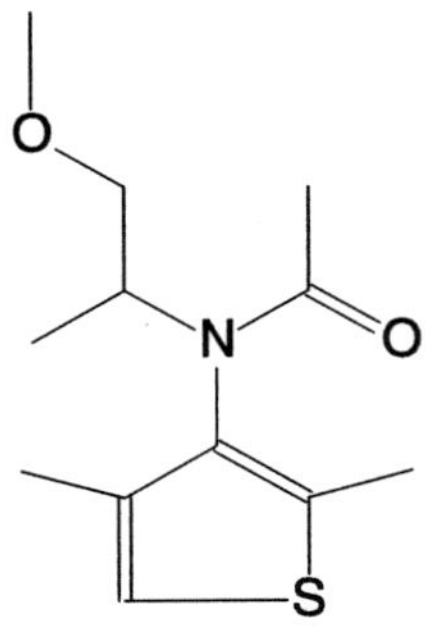

Figure 2.17 Structure of deschlorodimethenamid (XXX)

water was spiked with parent herbicides, neutral degradates, and the ionic degradates (i.e., the ESA and OA derivatives), using acetone (Ultraresi-analyzed from J.T. Baker; Phillipsburg, NJ) as a carrier solvent. Additional samples for recovery studies were prepared in deionized water that contained preservatives. The preservatives chosen were recommended by the U.S. EPA for drinking water analysis of organic compounds on the Contaminant Candidate List, including acetochlor (Winslow et al. 2001a; Winslow et al. 2001b). These consisted of ascorbic acid (0.57 mM; to reduce residual chlorine), ethylenediaminetetraacetic acid trisodium salt (0.98 mM; to chelate metal ions that might catalyze hydrolysis), diazolidinyl urea (3.6 mM; as a microbial inhibitor), and a pH 7 buffer of tris(hydroxymethyl)aminomethane and tris(hydroxymethyl)aminomethane hydrochloride (50 mM total). All preservatives were ACS reagent grade obtained from Aldrich. SPE recoveries of the ionic degradates were found in preliminary studies to be greatly suppressed by the presence of preservatives. Subsequent analyses of samples for ionic compounds were conducted by extracting unpreserved samples as soon as possible after collection. The recoveries for the DI water are shown in Table 2.1.

Table 2.1
Mean recoveries of parents and neutral degradates in triplicate 300 mL water samples, and of ionic degradates in triplicate 500 mL samples, all fortified at 3 μg/L. Note that preservatives were not used for the ionic degradates, as these were found to adversely affect recovery of these analytes.

		DI water		Surface water		Surface water 21 days	
No.	Compound	Mean recovery (%)	RSD* (%)	Mean recovery (%)	RSD (%)	Mean recovery (%)	RSD (%)
I	alachlor	96	7	93	3	93	2
II	hydroxyalachlor	92	8	106	5	102	4
III	deschloroalachlor	90	5	83	5	82	4
IV	2-chloro-2'-6'-diethylacetanilide	104	6	91	8	90	5
V	2-hydroxy-2'-6'-diethylacetanilide	90	4	93	2	92	4

(continued)

Table 2.1 (Continued)

No.	Compound	DI water Mean recovery (%)	DI water RSD[*] (%)	Surface water Mean recovery (%)	Surface water RSD (%)	Surface water 21 days Mean recovery (%)	Surface water 21 days RSD (%)
VI	2-hydroxy-2'-6'-diethyl-*N*-methylacetanilide	107	8	101	7	99	8
VII	2'-6'-diethylacetanilide	101	9	98	3	94	3
VIII	2,6-diethylaniline	92	5	91	4	89	6
IX	alachlor oxanilic acid	76	4	78	5	Nt[†]	
X	alachlor ethane sulfonic acid	93	5	94	4	Nt	
XI	metolachlor	100	4	97	4	97	4
XII	hydroxymetolachlor	107	7	104	4	102	8
XIII	deschlorometolachlor	100	2	88	5	89	2
XIV	metolachlor morpholinone	103	5	77	7	78	2
XV	metolachlor propanol	97	8	93	6	90	6
XVI	deschloroacetylmetolachlor	81	7	86	3	85	3
XVII	deschloroacetyl metolachlor propanol	104	3	87	3	86	3
XVIII	metolachlor oxanilic acid	85	3	87	5	Nt	
XIX	metolachlor ethane sulfonic acid	94	3	91	7	Nt	
XX	acetochlor	94	3	88	3	90	4
XXI	hydroxyacetochlor	102	6	107	4	101	3
XXII	deschloroacetochlor	89	6	86	3	87	4
XXIII	acetochlor oxanilic acid	81	4	80	4	Nt	
XXIV	acetochlor ethane sulfonic acid	96	5	98	4	Nt	
XXV	2-chloro-2'-ethyl-6'-methylacetanilide	105	4	88	3	89	1
XXVI	2-hydroxy-2'-ethyl-6'-methylacetanilide	83	4	95	6	91	9
XXVII	2'-ethyl-6'-methylacetanilide	88	3	93	1	91	3
XXVIII	2-ethyl-6-methylaniline	92	5	87	4	87	3
XXIX	dimethenamid	95	6	86	2	89	3
XXX	deschlorodimethenamid	96	5	83	1	85	6
XXXI	atrazine	99	6	83	6	86	6
XXXII	desethyl atrazine	82	10	81	5	83	1
XXXIII	desisopropyl atrazine	69	8	62	2	65	4
XXXIV	simazine	100	9	83	5	85	1
XXXV	cyanazine	103	8	88	4	87	4

* RSD = Relative standard deviation; n=3

† Nt = Not tested

Recoveries in Natural Waters

Additional studies were conducted to investigate whether the more complex matrix of natural waters could affect SPE recoveries, as well as to test the efficacy of the preservatives used in samples containing neutral analytes. Water used in SPE recovery and preservative studies was obtained from a local surface water impoundment, Loch Raven Reservoir (Towson, MD) on March 23, 2003. Samples were collected in two cleaned 4-L amber bottles, one containing preservatives and one without. The water sample containing preservatives was spiked with parent herbicides and neutral degradates. The second sample (without preservatives) was spiked with the ionic chloroacetamide degradates. Samples were then filtered through a 0.7 μm glass fiber filter (Millipore; Billerica, MA). Blank samples (no added herbicides or degradates) were also filtered in an identical manner prior to SPE. Half of the herbicide-spiked water containing preservatives was extracted immediately, and the other half was maintained at 4 °C for 21 days prior to SPE for re-analysis of parent compounds and neutral degradates. The samples without preservatives were extracted immediately. Results are shown in Table 2.1.

Sample Extraction and Derivatization

All solvents used in sample extraction and derivatization steps were Ultra Resi-Analyzed® from J.T. Baker. All extractions were performed using a Gilson (Middleton, WI) ASPEC XL solid phase extraction system. For the neutral analytes, extractions were performed with Oasis HLB (6 mL, 200 mg) cartridges from Waters (Milford, MA). Each cartridge was conditioned with 6 mL of methanol and 6 mL of deionized water. The samples (300 mL) were loaded onto the cartridge at 6 mL/min. After loading, the cartridges were again washed with 6 mL deionized water and 6 mL deionized water containing 15% methanol. The cartridges were not allowed to go to dryness during the sample loading or elution steps, although experiments revealed this did not affect recovery of our analytes on Oasis HLB SPE media. Analytes were eluted successively with 3 mL of methanol and 3 mL of ethyl acetate at a flow rate of 2 mL/min. The extracts were evaporated under a gentle stream of nitrogen to incipient dryness at ambient temperature. The final sample was brought up to 1 mL with toluene containing 50 μg/L of each of the three internal standards.

For the ionic compounds, an SPE procedure outlined by Shoemaker (2002) was followed. The SPE cartridges employed were Supelco Envi-Carb carbon (6 mL, 250 mg; Bellafonte, PA). Each cartridge was conditioned with 10 mL of 10 mM ammonium acetate in methanol and 25 mL of deionized water. The sample (500 mL) was loaded onto the cartridge at 6 mL/min. The cartridges were not allowed to go to dryness during the sample loading or elution steps. The cartridge was washed with 10 mL of deionized water. Ionic compounds were eluted with 6 mL of 10 mM ammonium acetate in methanol at a flow rate of 2 mL/min. The extracts were split into two 3-mL samples. Both samples were evaporated under a gentle stream of nitrogen to dryness at ambient temperature. The first sample, for HPLC-DAD analysis of the ESA degradates, was reconstituted in 0.5 mL of 10 mM ammonium acetate in deionized water containing 1 mg/L internal standard. The other sample, for GC/MS analysis of the methyl ester derivatives of the OAs, was reconstituted into 0.5 mL acetone for methylation using diazomethane. To the acetone fraction, an aliquot (200 μL-1 mL) of ether containing diazomethane was added until the yellow diazomethane color persisted for 15 minutes. The diazomethane was generated using a MNNG (*N*-methyl-*N*'-nitro-*N*-nitrosoguanidine)-diazomethane generator (Aldrich; note that diazomethane is explosive and

MNNG is mutagenic). The ether:acetone solvent was evaporated under nitrogen and the sample was brought to 0.5 mL volume with toluene containing 50 μg/L internal standards.

GC/MS Analysis

Injections of 1 μL (splitless) or 100 μL (large volume injections) were made onto a ThermoQuest (San Jose, CA) Trace 2000 gas chromatograph with a programmed temperature vaporization injector (PTV) coupled to a quadrupole mass spectrometer. A DB-35ms (Agilent; Palo Alto, CA) 30 m length × 0.25 mm ID × 0.25 μm phase thickness column was used to effect separations. The GC temperature program for the neutral compounds was 90 °C for 1 min, 6 °C/min to 290 °C, followed by a 5-min hold at 290 °C; the temperature program for the OA methyl esters was similar, except that the ramp rate was increased to 10 °C/min. For 1 μL splitless injections, the PTV injector was maintained at 200 °C. For 100 μL injections, the PTV was programmed to introduce the sample at 110 °C at a rate of 5 μL/sec. The evaporation occurred at a split flow rate of 100 mL/min for 0.5 min. The injector was programmed from 110 °C (0.5 min) to 290 °C (3.5 min) at 14 °C/sec with a splitless time of 2 min. The mass spectrometer temperature was set to 250 °C, with an energy of 70 eV, and spectra were obtained in electron ionization (EI) mode with selected ion monitoring, a *m/z* of 50-450 was scanned. The transfer line was maintained at 285 °C. Data were collected using Xcalibur software. Retention times for each compound and the quantitative and monitoring ions used are shown in Tables 2.2 and 2.3.

HPLC-DAD Analysis

The ESAs were analyzed via HPLC-DAD following a method similar to that developed by Hostetler and Thurman (2000). Injections of 100 μL were made onto a Waters HPLC-DAD (1525 pump and 2996 photodiode array detector) with Empower data acquisition system. The analytical wavelength was 210 nm. The mobile phase was 60:35:05 (10 mM ammonium acetate in water: methanol: acetonitrile) with a flow rate of 0.6 mL/min. A Phenomenex (Torrance, CA) Luna C_{18} 5 μm, 250 mm × 4.6 mm column was used to separate ESA degradates. Column temperature was set at 60 °C using a Phenomenex TS-130 column heater. While complete separation of the ESA degradates is achieved with this method, the metolachlor OA peak (if present in sufficiently high concentrations; >1 μg/L) overlaps with the acetochlor ESA and alachlor ESA peaks. In such a case this method cannot be used to quantify the concentrations of acetochlor ESA and alachlor ESA. Retention times are shown in Table 2.4.

Method Detection Limits

The method detection limits (MDLs) were obtained using the approach outlined in EPA Method 526 (Winslow et al. 2000). Instrument detection limits were obtained by injecting progressively more dilute aliquots of standard solutions until the relevant peaks could no longer be discerned from the background noise. For the neutral compounds, fourteen (seven for splitless and seven for LVI) 300-mL deionized water samples (with preservatives) were fortified at a concentration (10-700 ng/L for splitless; 0.10-7 ng/L for LVI) that would give a signal of 2 to 5 times the instrument detection limit after accounting for the 300-fold SPE concentration. For the ionic compounds, seven 500-mL deionized water samples (without preservatives) were fortified at a concentration (the OAs at 20 ng/L and the ESAs at 500 ng/L) that would give a signal of 2 to

Table 2.2
GC/MS data obtained for target analytes using selected ion monitoring (SIM)

Analyte No.	Compound	MW	t_R^* (min.)	Quantitation ion	Monitoring ion(s)
I	alachlor	269	21.55	160	237, 188
II	hydroxyalachlor	251	20.49	160	219, 188
III	deschloroalachlor	235	17.50	161	203, 178
IV	2-chloro-2'-6'-diethylacetanilide	225	18.38	176	225, 148
V	2-hydroxy-2'-6'-diethylacetanilide	207	21.1	176	207, 148
VI	2-hydroxy-2'-6'-diethyl-*N*-methylacetanilide	221	18.86	190	221, 162
VII	2'-6'-diethylacetanilide	191	16.64	148	191, 134
VIII	2,6-diethylaniline	149	10.31	134	149, 119
XI	metolachlor	283	22.72	162	238, 211
XII	hydroxymetolachlor	265	21.73	162	220, 193
XIII	deschlorometolachlor	249	18.62	162	204, 177
XIV	metolachlor morpholinone	233	21.05-21.15†	161	233, 188
XV	metolachlor propanol	269	24.57	162	238, 146
XVI	deschloroacetylmetolachlor	207	12.93	162	207, 133
XVII	deschloroacetyl metolachlor propanol	193	15.36	162	193, 133
XX	acetochlor	269	21.18	174	223, 162
XXI	hydroxyacetochlor	251	20.11	174	205, 146
XXII	deschloroacetochlor	235	17.04	164	206, 189
XXV	2-chloro-2'-ethyl-6'-methylacetanilide	211	17.39	162	211, 134
XXVI	2-hydroxy-2'-ethyl-6'-methylacetanilide	193	20.16	162	193, 134
XXVII	2'-ethyl-6'-methylacetanilide	177	15.59	120	177, 134
XXVIII	2-ethyl-6-methylaniline	135	8.88	120	135
XXIX	dimethenamid	275	21.25	154	230, 203
XXX	deschlorodimethenamid	241	17.11	154	196, 169
XXXI	atrazine	215	19.41	200	215, 173
XXXII	desethyl atrazine	187	18.14	172	187, 145
XXXIII	desisopropyl atrazine	173	18.31	173	158, 145
XXXIV	simazine	201	19.62	201	186, 173
XXXV	cyanazine	240	24.48	225	240, 212

* t_R= Retention time using DB-35 column, splitless injection

† Metolachlor morpholinone exits are four possible stereoisomers, only two of which could be separated on the GC/MS (given in the retention time range)

Table 2.3
GC/MS data for oxanilic acid methyl esters

No.	Compound	t_R^* (min.)	Quantitation ion	Monitoring ions
IX	Alachlor OA (methyl ester)	21.22	188	160, 146
XVII	Metolachlor OA (methyl ester)	22.27	248	188, 146
XXIII	Acetochlor OA (methyl ester)	20.81	174	162, 146

* t_R= Retention time using DB-35 column, splitless injection

Table 2.4
HPLC-DAD retention times for ethane sulfonic acids

No.	Compound	Retention time (min.)
X	Alachlor ESA	30.4
XVIII	Metolachlor ESA	34.2
XXIV	Acetochlor ESA	32.1

5 times the instrument detection limit after accounting for the 500-fold SPE concentration. These samples were extracted as described above. The standard deviation of the concentration of these samples was calculated. The method detection limit (MDL) = $S \cdot t_{(n-1,\ 1-\alpha=0.99)}$; where S = standard deviation of replicate analyses, $t_{(n-1,\ 1-\alpha=0.99)}$ = Students t value for the 99% confidence level with n-1 degrees of freedom, and n = number of replicates. The minimum reporting limit (MRL) was established as an analyte concentration that is three times the MDL.

A summary of the results is provided in Table 2.5. MDLs for the hydroxy-substituted compounds were almost two orders of magnitude larger than for the other compounds because of peak tailing and because of their extensive fragmentation on the MS. Fortunately we were able to use large volume injections (LVI) of 100 μL to decrease the MDLs for almost all of our neutral analytes by almost two orders of magnitude (MDLs = 4-100 ng/L for splitless injection and 0.06-4 ng/L for LVI). LVI did not prove useful for the two primary anilines (2-ethyl-6-methylaniline and 2,6-diethylaniline), which are too volatile for analysis via this technique. LVI was not pursued for the methyl ester derivatives of the OH degradates, which were present in all of our samples at concentrations sufficiently high enough to enable their analysis via GC/MS with 1 μL splitless injections.

DRINKING WATER UTILITY SAMPLES

Site Selection

Sites were chosen through contacts at American Water and several other agencies. This ensured a broad cross-section of water sources from both urban and rural areas. Final site selection emphasized a balance of surface and groundwater sources, as well as different treatment

Table 2.5
Method detection limits for target analytes analyzed via GC/MS or HPLC-DAD

No.	Compound	MDL* (ng/L) 1 µL	MDL (ng/L) 100 µL	MRL† (ng/L)
GC/MS				
I	alachlor	4	0.06	0.17
II	hydroxyalachlor	80	3	8.5
III	deschloroalachlor	6	0.2	0.71
IV	2-chloro-2'-6'-diethylacetanilide	10	0.1	0.40
V	2-hydroxy-2'-6'-diethylacetanilide	80	0.7	2.2
VI	2-hydroxy-2'-6'-diethyl-*N*-methylacetanilide	100	4	11
VII	2'-6'-diethylacetanilide	10	0.2	0.45
VIII	2,6-diethylaniline	10	Nt‡	33
IX	alachlor oxanilic acid	Nt	7	20
XI	metolachlor	5	0.1	0.31
XII	hydroxymetolachlor	70	1	3.3
XIII	deschlorometolachlor	9	0.2	0.55
XIV	metolachlor morpholinone	10	0.2	0.45
XV	metolachlor propanol	10	0.2	0.54
XVI	deschloroacetylmetolachlor	9	0.1	0.29
XVII	deschloroacetyl metolachlor propanol	80	0.8	2.5
XVIII	metolachlor oxanilic acid	Nt	7	20
XX	acetochlor	8	0.2	0.44
XXI	hydroxyacetochlor	60	0.2	12
XXII	deschloroacetochlor	9	0.07	0.20
XXIII	acetochlor oxanilic acid	Nt	7	20
XXV	2-chloro-2'-ethyl-6'-methylacetanilide	10	0.2	0.65
XXVI	2-hydroxy-2'-ethyl-6'-methylacetanilide	80	0.2	2.4
XXVII	2'-ethyl-6'-methylacetanilide	10	0.2	0.58
XXVIII	2-ethyl-6-methylaniline	8	Nt	25
XXIX	dimethenamid	7	0.1	0.29
XXX	deschlorodimethenamid	9	0.1	0.30
XXXI	atrazine	5	0.2	0.55
XXXII	desethyl atrazine	10	0.3	0.77
XXXIII	desisopropyl atrazine	20	0.2	0.44
XXXIV	simazine	10	4	0.61
XXXV	cyanazine	10	0.1	0.34
HPLC-DAD				
X	alachlor ethane sulfonic acid	Nt	100	320
XIX	metolachlor ethane sulfonic acid	Nt	90	280
XXIV	acetochlor ethane sulfonic acid	Nt	100	320

* MDLs were computed by analyzing seven replicate samples near the detection limit and multiplying the SD by t (for 99% confidence interval, n-1)
† MRLs are 3 times the MDL
‡ Nt = Not tested

Table 2.6
Selected drinking water utility sites and some of their characteristics (further information about the treatment train employed at each site can be found in Appendix A)

State	Source water	Ozone	GAC/PAC	Flow (MGD)
IA	Groundwater	No	No	35
IA	Surface water	No	Yes	16
IL	Surface water	No	Yes	28
IL	Surface water/groundwater	No	Yes	11
IL	Surface water	No l	No	2.6
MO	Groundwater	No	No	14
MO	Surface water	No	Yes (not in fall)	4.4
MO	Groundwater	Yes	No	2.4
IN	Surface water/groundwater	No	Yes	10
IN	Surface water	No	Yes (not in fall)	3.5
KY	Surface water	No	No	95
KY	Surface water	No	No	37

trains (Table 2.6). More detailed information pertaining to treatment trains is provided in Appendix A. The one site that previously employed riverbank filtration did not do so during our sampling period. The wells for the riverbank-filtered water required repairs and were out of service during the sampling period. The questionnaire filled out by each utility at the time of sampling can be found in Appendix B.

Drinking Water Sample Collection

The fall water samples were collected between October 13 and December 11, 2003. The spring water samples were collected between June 1 and June 17, 2004. Samples were collected in I-Chem 200 series 2 L amber bottles (Fisher Scientific; Fairlawn, NJ), to which preservatives/pH buffer had been added by us, as appropriate; these were prepared in our laboratory and were then shipped to the drinking water utility personnel. No preservatives were added to samples for analysis of the ethane sulfonic acid (ESA) or oxanilic acid (OA) degradates, as they interfere with solid phase extraction of these analytes. At each plant, two 2-L bottles of raw water and two 2-L bottles of finished water were collected as grab samples. All field samples were identified by metal tags attached to the handle of the bottle marked in ink indicating site number, location, date, and time of sampling. Sampling protocols were provided (in writing) to water treatment plant personnel collecting the samples. A copy of the sampling protocol can be found in Appendix B. The samples were shipped overnight in coolers without ice packs to Johns Hopkins University. A summary and schematic of the sample division are described in Table 2.7 and Figure 2.18.

Table 2.7
Division of samples for each drinking water treatment facility (raw and treated water) prior to SPE

	Sample size	GC/MS or HPLC samples	Laboratory fortified sample matrix	Storage (10°C)
Neutral compounds	2 L	3 × 300 mL	1 × 300 mL	2 × 300 mL
Ionic compounds	2 L	2 × 500 mL	1 × 500 mL	—

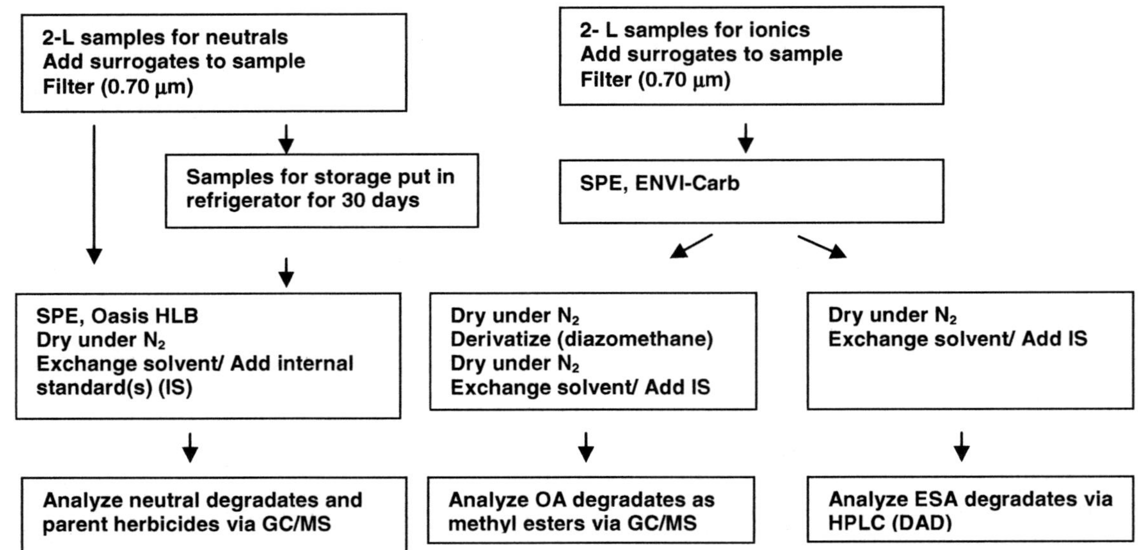

Figure 2.18 Schematic of analytical procedure

Sample bottles for the neutral compounds (one for raw and one for finished water) contained 0.10 g/L ascorbic acid (to reduce residual chlorine), 0.35 g/L ethylenediamine-tetraacetic acid trisodium salt (to chelate metal ions), 1.0 g/L diazolidinyl urea (as a microbial inhibitor), 0.47 g/L tris(hydroxymethyl)aminomethane and 7.28 g/L tris(hydroxymethyl) aminomethane hydrochloride (to buffer the sample to pH 7). The preservatives chosen were recommended by the U.S. EPA for drinking water analysis of organic compounds on the CCL, including acetochlor (Winslow et al. 2001a; Winslow et al. 2001b). All preservatives were ACS reagent grade obtained from Aldrich.

At the time of receipt, subsamples from each bottle containing preservatives were tested to verify that the preservatives were functioning correctly. Microbial growth was assessed via a heterotrophic plate count method (with subsamples from the preservative-free bottles as the controls; APHA, AWWA, and WEF 1992). Residual chlorine was tested using a DPD colorimetric method (APHA, AWWA, and WEF 1992).

Quality Control

Upon receipt by our laboratory, surrogate standards were added to the samples at concentrations of 100 ng/L (for neutral surrogates) and 500 ng/L (for the ionic surrogate standard). From each 2-L bottle for the neutral compounds, the water was divided accordingly: three 300-mL samples were extracted for neutral compounds and one 300-mL sample was fortified with the neutral analytes (200 ng/L). From each 2-L bottle for the ionic degradates, the water was divided as follows: two 500-mL samples were extracted for ionic compounds, and one 500-mL sample was fortified with the ionic analytes (500 ng/L). The water samples were filtered with a 0.7 μm glass fiber filter (Millipore; Billerica, MA) after addition of surrogates or fortification but prior to extraction.

Additional quality control was maintained by analyzing laboratory blanks and laboratory fortified blanks according to procedures outlined in U.S. EPA Method 526 (Winslow et al. 2000). Laboratory blanks consisted of deionized water. Laboratory fortified blanks contained deionized water spiked with neutral (5-500 ng/L) or ionic (100-500 ng/L) analytes. The laboratory blanks, laboratory fortified blanks and laboratory fortified field samples were handled in the same manner as the field samples.

BENCH SCALE TREATMENT TESTS

Coagulation

Optimal Coagulant Dose

Jar tests were carried out following procedures given by Hudson (1981) using water from Loch Raven Reservoir (which supplies drinking water to the Baltimore, MD metropolitan area) to determine the appropriate coagulant dose. Alum (aluminum sulfate, $Al_2(SO_4)_3 \cdot 18H_2O$; Aldrich) and ferric chloride ($FeCl_3$; Aldrich) were dosed at 0, 10, 20, 30, 40 and 50 mg/L. Six 1-L beakers filled with Loch Raven Reservoir water were placed in a laboratory stirrer (Phipps and Bird; Richmond, VA). Coagulants (as a 1.5% by weight dosing solution) were added via pipette to yield the appropriate concentrations. Samples underwent 2 min of rapid mixing at 100 rpm, followed by 60 min of slow mixing at 20 rpm. After the completion of the slow mix period, the samples were allowed to settle for one hour. Supernatant samples were collected for measurement of turbidity (HF Scientific DRT 100B; Ft. Meyers, FL), UV-absorbance at 254 nm (Shimadzu UV-1601; Columbia, MD) and TOC concentration (Dohrmann Phoenix 8000 UV-persulfate TOC analyzer; Mason, OH). The optimum coagulant dose was chosen as the dose at which turbidity after settling was minimized.

Reductions in TOC concentrations following coagulation were checked for compliance with the Enhanced Coagulation Rule (White et al. 1997). The Enhanced Coagulation Rule is a regulatory strategy to limit the formation of disinfection by-products through the removal of TOC during coagulation. For the Loch Raven water under study, which has a TOC of ~3 mg/L, a 40% reduction in the TOC upon coagulation is required. Addition of alum and ferric chloride met the enhanced coagulation requirement.

Compound Removal

The compounds of interest were spiked into Loch Raven water at 50 µg/L (in groups of three to four compounds). Each compound group was tested in triplicate with three 1-L beakers for each coagulant. The coagulant was added at the optimum dose as determined from the previous step (alum, 30 mg/L; $FeCl_3$, 20 mg/L). Upon settling, aliquots were removed to determine remaining herbicide concentrations as described below.

Oxidation

Chlorination

The free chlorine dosing solution was prepared by diluting a ~ 4-6% aqueous solution of NaOCl (Fisher Scientific; Fairlawn, NJ) to a concentration of approximately 200 mg/L free chlorine and adjusting the pH to 7. This solution was standardized iodometrically (APHA, AWWA, and WEF 1992). Serum bottles (60 mL) were prepared containing Loch Raven Reservoir water, 50 µg/L of the test compound and 6 mg/L free chlorine. Test compounds were added as an acetone solution at acetone volumes ≤ 10 µL. Each compound was treated individually in triplicate bottles. The bottles were closed with Teflon lined rubber stoppers with crimp caps and were mixed (after dosing) on a stir plate for one minute. The bottles were then incubated for six hours at 22 °C ± 1 °C in the dark. At the end of the reaction period, the samples were immediately extracted for analysis as described below, eliminating the need for a chlorine quenching agent. One of the three bottles was also tested for residual chlorine; all samples tested had a residual free chlorine concentration >1 mg/L via a DPD colorimetric method (APHA, AWWA, and WEF 1992).

Ozonation

Stock ozone dosing solutions were prepared according to Standard Method 2350 (APHA, AWWA, and WEF 1992) by bubbling an ozone/oxygen gas stream (OREC V5-O ozone generator; Phoenix, AZ) through Milli-Q water (Millipore; Billerica, MA) placed on an ice bath. Water samples (800 mL) were ozonated for one hour with continuous stirring. The final ozone concentration in the water was measured via an indigo colorimetric method with a HACH (Loveland, CO) DR-100 colorimeter. With this protocol, typical ozone concentrations in the aqueous dosing solution were 24 to 26 mg/L.

Reactivity to ozone was studied in triplicate 60 mL serum vials that contained Loch Raven Reservoir water and a single test compound (50 µg/L). Test compounds were added as an acetone solution at acetone volumes ≤ 10 µL. The applied ozone dosage (3 mg/L) was approximately 1 mg of ozone per mg of TOC in the sample. The bottles were prepared without headspace and were sealed with Teflon-lined rubber stoppers and crimp caps. The bottles were stirred and incubated for 30 min in the dark at 22 °C ± 1 °C. At the end of the reaction period, the samples were immediately extracted as described below for analysis of the test compound; ozone quenching agents were not used. The ozone level was not measured upon completion of the reaction.

Adsorption

Calgon (Pittsburgh, PA) Water Powdered High Activity (WPH) powdered activated carbon (PAC) was used for these experiments. The PAC was dried in an oven for three hours at 150 °C before use. Adsorption isotherms were determined at room temperature (22 °C ± 1 °C) in a series of five 60-mL serum bottles with Teflon-lined crimp caps. Bottles contained filtered (0.7 µm; Millipore) Loch Raven Reservoir water, 50 µg/L of each individual test compound, and a known mass of PAC (prepared by diluting a PAC slurry in deionized water). A control sample was run in the absence of PAC and did not show any loss of the test compound. The PAC concentrations ranged from 0.5 to 8 mg/L. Bottles were covered in aluminum foil and were placed on a roller table at 45 rpm for 5 days. In separate 10-day kinetic studies, a contact period of 5 days was found sufficient to attain 99% of the long-term value. After 5 days, the bottle contents were filtered through a 0.45 µm nylon membrane filter (Millipore) and were extracted as described below for analysis of the test compound.

Quantitative Analysis for Treatment Trains

GC-NPD

Compounds analyzed via this method were: alachlor (**I**), deschloroalachlor (**III**), 2-chloro-2'-6'-diethylacetanilide (**IV**), 2'-6'-diethylacetanilide (**VII**), 2,6-diethylaniline (**VIII**), metolachlor (**IX**), deschlorometolachlor (**XI**), deschloroacetylmetolachlor (**XIV**), acetochlor (**XVI**), deschloroacetochlor (**XVIII**), 2-chloro-2'-ethyl-6'-methylacetanilide (**XIX**), 2'-ethyl-6'-methylacetanilide (**XXI**), 2-ethyl-6-methylaniline (**XXII**), dimethenamid (**XXIII**), deschlorodimethenamid (**XXIV**) and atrazine (**XXV**).

Three 10-mL aliquots of each sample were successively extracted with 1 mL of dichloromethane for 1 min; the three 1-mL extracts were then combined. The dichloromethane was dried under a gentle stream of nitrogen and the sample was reconstituted in 0.50 mL of toluene containing 2-nitro-*m*-xylene as the internal standard. Injections (1 µL) were made on-column onto a Carlo Erba (San Jose, CA) Mega 2 GC with a flameless nitrogen phosphorus detector (NPD). A DB-5 (Agilent; Palo Alto, CA) 30 m length × 0.25 mm ID × 0.25 µm film thickness column was used to effect separations. The GC temperature program was 110 °C for 1 min, 10 °C/min to 290 °C, followed by a 5-min hold at 290 °C. Extraction efficiencies ranged from 96-100%.

HPLC-DAD

Compounds analyzed via this method were: hydroxyalachlor (**II**), 2-hydroxy-2'-6'-diethylacetanilide (**V**), 2-hydroxy-2'-6'-diethyl-*N*-methylacetanilide (**VI**), hydroxymetolachlor (**X**), metolachlor morpholinone (**XII**), metolachlor propanol (**XIII**), deschloroacetylmetolachlor propanol (**XV**), hydroxyacetochlor (**XVII**) and 2-hydroxy-2'-ethyl-6'-methylacetanilide (**XX**).

Three 10-mL aliquots of each sample were successively extracted with 1 mL of dichloromethane for 1 min; the three 1-mL extracts were then combined. The dichloromethane was dried under a gentle stream of nitrogen and the sample was reconstituted in 0.50 mL of 10 mM ammonium acetate in acetonitrile containing 2,4-dichlorophenylacetic acid as the internal standard. A 100-µL sample of each extract was then injected onto a Waters (Milford, MA) HPLC-DAD system (1525 pump and 2996 photodiode array detector), with the detector set to an analytical wavelength of 210 nm. The mobile phase was 50% 10 mM aqueous ammonium acetate and 50% acetonitrile, with a flow rate of 1.0 mL/min. The analytical column was a Phenomenex (Torrance, CA) Luna C_{18} 5 μm, 250 mm × 4.6 mm. The column temperature was set at 60 °C using a Phenomenex TS-130 column heater. Extraction efficiencies ranged from 93-98%.

Product Identification (GC/MS)

Injections of 1 μL were made onto a ThermoQuest (San Jose, CA) Trace 2000 gas chromatograph with a programmed temperature vaporization injector (PTV) coupled to a quadrupole mass spectrometer. A DB-5ms (Agilent) 30 m length 0.25 mm ID × 0.25 μm film thickness column was used to effect separations. The GC temperature program was 90 °C for 1 min, 10 °C/min to 290 °C, followed by a 5-min hold at 290 °C. The PTV injector was maintained at 200 °C. The mass spectrometer temperature was set to 250 °C, with an energy of 70 eV, and spectra were obtained in electron ionization (EI) mode. The transfer line was maintained at 285 °C.

K_{ow} Estimation

Estimates of the log K_{ow} value for each parent compound and neutral degradate were determined using ClogP version 2.0 (BioByte; Claremont, CA).

CHAPTER 3
RESULTS AND DISCUSSION

FALL 2003 DRINKING WATER SAMPLES

Herbicide and Herbicide Degradates

The sites, their assigned numbers, date sampled and treatment used at the time of sampling can be found in Table 3.1. A summary of the overall occurrence, maximum and median concentrations can be found in Table 3.2. All parent herbicides were detected, and 25 of 28 degradates were detected in at least one sample. Most of the compounds were found in at least half of the raw and finished samples. Atrazine was the parent herbicide with the highest maximum and median concentration, followed by metolachlor. Of the degradates, the ionic compounds had the highest maximum and median concentrations, about approximately one order of magnitude higher than either the parent herbicides or the neutral degradates (~100-1000 ng/L versus ~10-100 ng/L). The neutral degradates tended to be found in concentrations similar to that of the parent compound, ~10-100 ng/L. Overall, no significant removal (>50%) was observed for the parent herbicides or their degradates despite the variety of water treatment processes employed. Details concerning measured concentrations of each compound in the raw and finished water, standard deviations and removal percentages for the individual sites can be found in Tables A.1 through A.12 in Appendix A.

Quality Control

Recoveries were >90% for the neutral surrogates, $^{13}C_6$-metolachlor and $^{13}C_3$-atrazine, while the recovery for the ionic surrogate 2-benzoylbenzoic acid was >60% for all of the samples. Recoveries of all the fortified sample matrices and fortified blanks were >70% for all of the compounds except for desisopropyl atrazine, for which recovery was >65%. None of the blanks revealed interference for any of the compounds of interest.

The DPD tests showed no residual chlorine in any of the samples, indicating the ascorbic acid was functioning properly. Heterotrophic plate counts for all of the samples containing preservatives revealed substantial decreases in the microbial growth with respect to the samples that did not have preservatives.

Storage Samples

The aliquots left for storage were maintained in their original amber bottles and were placed in cold storage at 6°C for 30 days. Half of the storage samples were analyzed, while the other half were discarded. Of the sites analyzed for storage effects, all of the compounds measured were within one standard deviation of the concentrations measured prior to storage. Details concerning measured concentrations for the individual sites can be found in Tables A.13 through A.18.

Table 3.1
Data from sites sampled during Fall 2003

Site no.	Date sampled (mm/dd/yy)	Source	Treatment
1	10/13/03	Surface water	Lime softener, $FeCl_3$, carbonation, chlorine
2	11/05/03	Surface water	$KMnO_4$, cationic polymer, poly-Al sulfate, anionic polymer, chlorine, GAC/ sand filter, ammonia, caustic soda, fluoride
3	10/22/03	Surface water	PAC, chlorine, ammonia, coagulant (Clarion 410p), GAC filters, zinc ortho, caustic soda, fluoride
4	11/03/03	Surface water	Alum, cationic polymer
5	11/23/03	Surface water	Cationic polymer, lime, $FeCl_3$, ortho phosphate, chlorine, fluoride
6	11/03/03	Surface water/ groundwater	Coagulant, multi-media filters with GAC
7	10/30/03	Surface water/ groundwater	$KMnO_4$, caustic soda, chlorine, coagulant, GAC filters, chlorine, phosphate, ammonia, fluoride
8	11/03/03	Groundwater	Ozone, chloride, poly-phosphate
9	11/13/03	Groundwater	Aeration, lime, chlorine, filtration with dual media filters (no GAC or PAC)
10	12/11/03	Groundwater	Softening, settling, chlorine, CO_2, filtration, corrosion inhibitor, fluoride
11	11/24/03	Surface water	$FeCl_3$, chlorine, ammonia, quicklime, aluminum chlorohydrate, anthracite sand filters
12	11/24/03	Surface water/RBF	$FeCl_3$, quicklime, chlorine, filters, ammonia

SPRING 2004 DRINKING WATER SAMPLES

Herbicide and Herbicide Degradates

The sites, their assigned numbers, date sampled and treatment used at the time of sampling can be found in Table 3.3. A summary of the overall occurrence and maximum and median concentrations can be found in Table 3.4. All parent herbicides were detected, and 27 of 28 degradates were detected in at least one sample. Most of the compounds were found in at least half of the raw and finished samples. Atrazine was the parent herbicide with the highest maximum and median concentration (~2 μg/L), followed by metolachlor and acetochlor (~200-300 ng/L). Simazine and dimethenamid were approximately an order of magnitude lower in concentration (10-30 ng/L) than acetochlor or metolachlor. Alachlor was detected in almost all of the samples but at a concentration around ~5 ng/L. Cyanazine was encountered in less than a quarter of the samples and at low concentrations (~10 ng/L). The low concentrations of alachlor and cyanazine most likely reflect their

Table 3.2
Percent detection, maximum and median concentration for raw and finished drinking water samples obtained in Fall 2003

No.	Compound	n	% Detection Raw	% Detection Finished	Maximum concentration (ng/L) Raw	Maximum concentration (ng/L) Finished	Median concentration (ng//L) Raw	Median concentration (ng//L) Finished
I	alachlor	12	67	67	17	17	7.4	6.6
II	hydroxyalachlor	12	33	25	46	44	23	25
III	deschloroalachlor	12	0	0				
IV	2-chloro-2'-6'-diethylacetanilide	12	100	100	15	11	6.3	6.4
V	2-hydroxy-2'-6'-diethylacetanilide	12	25	25	16	15	11	9.2
VI	2-hydroxy-2'-6'-diethyl-*N*-methylacetanilide	12	0	0				
VII	2'-6'-diethylacetanilide	12	100	100	43	38	29	26
VIII	2,6-diethylaniline	12	50	42	<11	<11	<11	<11
IX	alachlor oxanilic acid	12	75	75	68	69	58	54
X	alachlor ethane sulfonic acid	12	17	17	945	743	673	552
XI	metolachlor	12	100	100	51	49	11	11
XII	hydroxymetolachlor	12	75	75	49	35	29	24
XIII	deschlorometolachlor	12	100	100	6.0	6.3	3.3	3.3
XIV	metolachlor morpholinone	12	17	8	34	8.8	23	8.8
XV	metolachlor propanol	12	0	8		15		15
XVI	deschloroacetylmetolachlor	12	75	83	5.1	5.6	2	2
XVII	deschloroacetyl metolachlor propanol	12	0	0				
XVIII	metolachlor oxanilic acid	12	92	92	138	137	61	58
XIX	metolachlor ethane sulfonic acid	12	67	67	1063	1010	441	420
XX	acetochlor	12	50	50	56	45	5.6	4.5
XXI	hydroxyacetochlor	12	50	50	60	55	36	31
XXII	deschloroacetochlor	12	33	42	27	20	14	13
XXIII	acetochlor oxanilic acid	12	100	100	86	82	57	54
XXIV	acetochlor ethane sulfonic acid	12	25	25	645	574	560	533
XXV	2-chloro-2'-ethyl-6'-methylacetanilide	12	100	100	167	163	21	20
XXVI	2-hydroxy-2'-ethyl-6'-methylacetanilide	12	42	33	33	35	4.0	13
XXVII	2-ethyl-6-methylaniline	12	92	92	<25	<25	<25	<25
XXVIII	2'-ethyl-6'-methylacetanilide	12	100	100	57	57	38	38
XXIX	dimethenamid	12	58	50	8.2	7.7	3.4	2.5
XXX	deschlorodimethenamid	12	67	67	14	25	5.3	3.7
XXXI	atrazine	12	100	100	266	225	84	77
XXXII	desethyl atrazine	12	92	92	56	46	20	21
XXXIII	desisopropyl atrazine	12	50	58	21	33	10	10
XXXIV	simazine	12	92	92	49	32	6.6	6.3
XXXV	cyanazine	12	50	50	12	12	10	9.4

Table 3.3
Data from sites sampled during Spring 2004

Site no.	Date sampled (mm/dd/yy)	Source	Treatment
1	06/03/2004	Surface water	Cationic polymer, lime, $FeCl_3$, PAC, CO_2-recarbonation, chlorine, fluoride, phosphate, CO_2, chlorine, GAC capped sand filters, ammonium sulfate, chlorine
2	06/02/2004	Surface water	Chlorine, $KMnO_4$, cationic polymer, poly-aluminum, sulfate, anionic polymer, chlorine, GAC, chlorine, ammonia, caustic soda, fluoride, zinc ortho
3	06/07/2004	Surface water	PAC, free chlorine (chloramines), ammonia, Clarion 410p coagulant, filter (with GAC), zinc ortho, chlorine, ammonia, fluoride
4	06/15/2004	Surface water	Alum, cationic polymer, chlorine, ammonia
5	06/14/2004	Surface water	Polymer catfloc polydyne, lime, $FeCl_3$, GAC filters, orthophosphate, chlorine, fluoride
6	06/17/2004	Surface water/ groundwater	Polyaluminum hydroxychloride/polyquatenaryamine blend, PAC, $KMnO_4$, copper sulfate, GAC filters, chlorine
7	06/14/2004	Surface water/ groundwater	PAC, chlorine, coagulant/coagulant aid, GAC, chlorine, ammonia, phosphate, fluoride, caustic soda
8	06/02/2004	Groundwater	Ozone, chlorine, polyphosphate
9	06/14/2004	Groundwater	Lime softening, aeration, lime, chlorine, fluoride, filtration (dual media rapid sand)
10	06/04/2004	Groundwater	Ammonia, aeration, lime, chloride, CO_2, filtration, fluoride, phosphate
11	06/01/2004	Surface water	$FeCl_3$, Poly (Diallyldimethylammonium Chloride) (pDADMAC), chlorine, ammonia, lime, aluminumchlorohydrate, anthracite/sand filtration
12	06/01/2004	Surface water	$FeCl_3$, pDADMAC, chlorine, ammonia, lime

decreasing use on agricultural crops. Of the degradates, the ionic compounds (oxanilic acids and ethane sulfonic acids) had the highest maximum and median concentrations, and were more abundant (~200-900 ng/L) than the parent chloroacetamide herbicides. The neutral degradates tended to be found at concentrations lower than the parent herbicides (~10-100 ng/L for the neutral chloroacetamide degradates and ~100-300 ng/L for the atrazine degradates).

Significant removal of parent herbicides and degradates from the raw to the finished water was encountered in those plants that employed activated carbon. The type of activated carbon did not seem to have a substantial effect on removal; facilities with granular, powdered, or a combination of the two experienced average removals of approximately 40% for all compounds under investigation. Ozonation was used at one treatment facility. Removal of the parent herbicides at this facility ranged from 30 to 100%. While some neutral degradates were removed on ozonation (up to

Table 3.4
Percent detection and maximum and median concentrations for raw and finished drinking water samples obtained in Spring 2004

			% Detection		Maximum concentration (ng/L)		Median concentration (ng//L)	
No.	Compound	n	Raw	Finished	Raw	Finished	Raw	Finished
I	alachlor	12	92	83	11	5.6	3.5	1.6
II	hydroxyalachlor	12	83	67	43	34	28	28
III	deschloroalachlor	12	17	8	14	0.7	8.7	0.7
IV	2-chloro-2'-6'-diethylacetanilide	12	100	100	3.7	3.4	1.1	1.2
V	2-hydroxy-2'-6'-diethylacetanilide	12	100	100	104	85	55	28
VI	2-hydroxy-2'-6'-diethyl-*N*-methylacetanilide	12	50	42	1.7	1.7	1.7	1.7
VII	2'-6'-diethylacetanilide	12	100	100	13	15	6.8	6.3
VIII	2,6-diethylaniline	12	0	0	Nd	Nd	Nd	Nd
IX	alachlor oxanilic acid	12	92	92	216	136	110	75
X	alachlor ethane sulfonic acid	12	17	17	648	618	554	460
XI	metolachlor	12	100	100	891	311	318	157
XII	hydroxymetolachlor	12	75	75	217	61	41	30
XIII	deschlorometolachlor	12	100	100	32	30	5.8	5.0
XIV	metolachlor morpholinone	12	100	92	63	37	41	27
XV	metolachlor propanol	12	100	92	208	73	35	28
XVI	deschloroacetylmetolachlor	12	58	67	39	35	8.7	5.1
XVII	deschloroacetyl metolachlor propanol	12	25	25	17	22	15	13
XVIII	metolachlor oxanilic acid	12	92	92	687	215	158	113
XIX	metolachlor ethane sulfonic acid	12	92	92	1580	1530	888	594
XX	acetochlor	12	100	100	991	372	225	120
XXI	hydroxyacetochlor	12	92	83	198	64	87	56
XXII	deschloroacetochlor	12	92	92	35	31	11	5.6
XXIII	acetochlor oxanilic acid	12	92	92	1170	551	326	231
XXIV	acetochlor ethane sulfonic acid	12	58	58	1080	845	870	557
XXV	2-chloro-2'-ethyl-6'-methylacetanilide	12	100	92	33	7.0	10	3.9
XXVI	2-hydroxy-2'-ethyl-6'-methylacetanilide	12	92	92	105	67	50	21
XXVII	2'-ethyl-6'-methylacetanilide	12	100	100	25	17	9.3	9.0
XXVIII	2-ethyl-6-methylaniline	12	0	8	Nd	<25	Nd	Nd
XXIX	dimethenamid	12	83	75	308	67	13	3.3
XXX	deschlorodimethenamid	12	50	33	1.8	1.3	1.4	1.2
XXXI	atrazine	12	100	100	4250	1860	1560	780
XXXII	desethyl atrazine	12	100	100	594	318	266	142
XXXIII	desisopropyl atrazine	12	100	100	199	75	92	39
XXXIV	simazine	12	100	100	201	193	27	15
XXXV	cyanazine	12	25	17	19	11	9.6	7.2

100%), other neutral degradates were apparently formed during such treatment, resulting in concentrations up to twice that in the raw water. Those facilities that did not use activated carbon or ozone encountered little removal of any of the compounds under investigation. Details concerning measured concentrations of each compound in the raw and finished water, standard deviations and removal percentages for the individual sites can be found in Tables A.1 through A.12.

When the samples obtained during the spring of 2004 are compared with those obtained in fall of 2003, the parent herbicides are present at concentrations approximately an order of magnitude higher in the spring samples. This was expected, as these herbicides are applied in early spring. The ionic degradate concentrations were also present at higher concentrations in the spring than in the fall. The neutral degradates were detected at higher concentrations in the spring, but the difference between the spring and fall samples was not as great as it was for the parent compounds. More in-depth comparisons of the removal data can be found elsewhere (Hladik 2005).

Quality Control

Recoveries were >90% for the neutral surrogates, $^{13}C_6$-metolachlor and $^{13}C_3$-atrazine, while the recovery for the ionic surrogate 2-benzoylbenzoic acid was >70% for all of the samples. Recoveries of all the fortified sample matrices and fortified blanks were >70% for all of the compounds except for desisopropyl atrazine, for which recovery was >65%. None of the blanks revealed interference for any of the compounds of interest.

The DPD tests did not reveal residual chlorine in any of the samples, indicating that the ascorbic acid was functioning properly as a reductant. Heterotrophic plate counts for all of the samples containing preservatives revealed substantial decreases in microbial growth with respect to the samples that did not contain preservatives.

COMPARISON OF RESULTS TO DRINKING WATER QUALITY CRITERIA

Data pertaining to measured concentrations in the finished water at all twelve sites sampled are shown in Figures 3.1 through 3.12. U.S. drinking water quality criteria for alachlor (2 μg/L) and atrazine (2 μg/L) are met at all the sites investigated on the days sampled; acetochlor and metolachlor are not currently regulated (EPA 2001b). Bar charts reveal the contribution of the parent herbicides, the neutral degradates and the ionic degradates to the total amount present in water samples. In these charts, those degradates that can form from either metolachlor or acetochlor were apportioned according to the concentrations of metolachlor and acetochlor found at the site in question (on a percentage basis of the molar concentration).

As acetochlor and metolachlor, as well as chloroacetamide degradates, are on the U.S. EPA's CCL list, it might be instructive to speculate whether water quality criteria would be likely to have been exceeded if the MCL for alachlor were to be redefined at the same value, only including its neutral or ionic degradates. None of the finished water samples would have exceeded a hypothetical MCL of 2 μg/L for alachlor even if both the neutral and ionic degradates were included.

It may also prove instructive to speculate as to whether the sites analyzed in this study would have met drinking water quality criteria if MCL values for acetochlor or metolachlor were to be established that are identical to the current MCL for alachlor. An MCL value for acetochlor of 2 μg/L might not seem unreasonable given their close similarity in toxicity (Dearfield et al. 1999) as well as the structure. An argument could be made for a somewhat higher MCL for metolachlor, because of differences in toxicity (Dearfield et al. 1999) and reactivity (Carlson

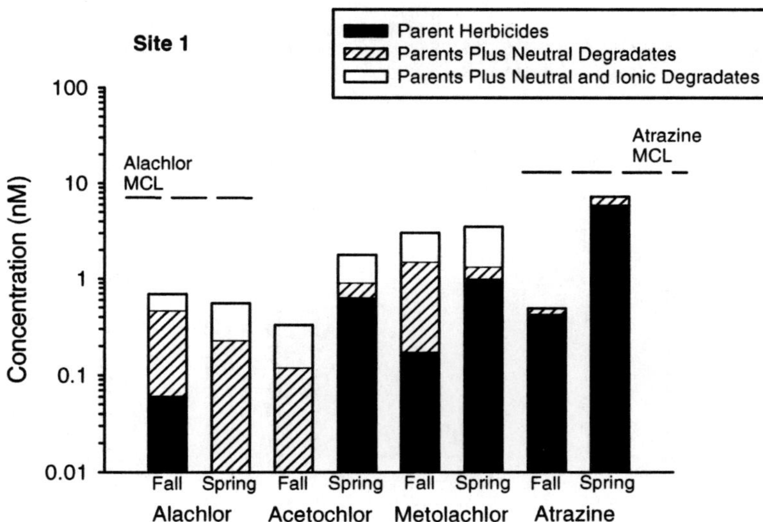

Figure 3.1 Bar chart showing concentrations (on a molar basis) of the parent chloroacetamide and chloro-*s*-triazine herbicides and degradates measured at Site 1 in Fall 2003 and Spring 2004 in the finished water. Those degradates that could originate from either acetochlor or metolachlor were apportioned according to the relative abundance of the parent herbicides at this site. Dashed lines represent the U.S. EPA's maximum contaminant level (MCL) for alachlor and atrazine.

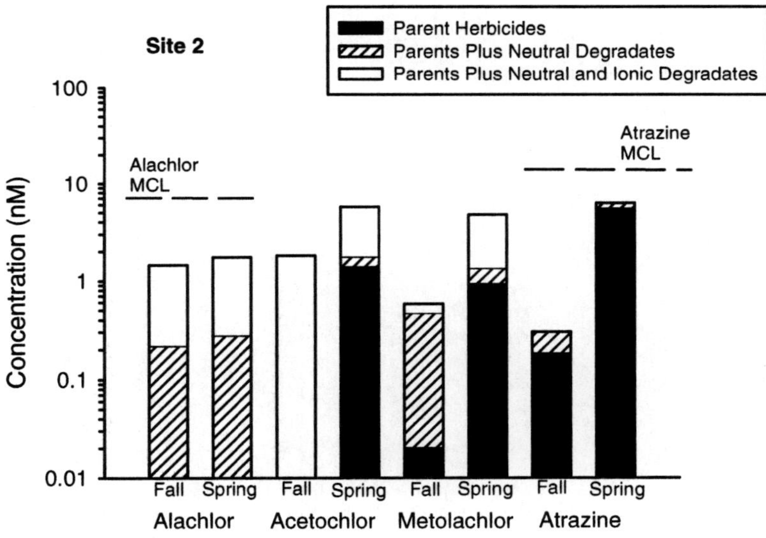

Figure 3.2 Bar chart showing concentrations (on a molar basis) of the parent chloroacetamide and chloro-*s*-triazine herbicides and degradates measured at Site 2 in Fall 2003 and Spring 2004 in the finished water. Those degradates that could originate from either acetochlor or metolachlor were apportioned according to the relative abundance of the parent herbicides at this site. Dashed lines represent the U.S. EPA's maximum contaminant level (MCL) for alachlor and atrazine.

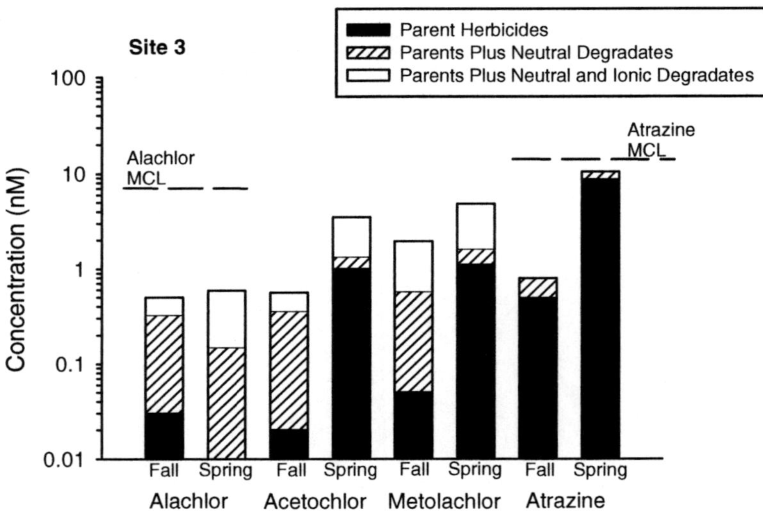

Figure 3.3 Bar chart showing concentrations (on a molar basis) of the parent chloroacetamide and chloro-*s*-triazine herbicides and degradates measured at Site 3 in Fall 2003 and Spring 2004 in the finished water. Those degradates that could originate from either acetochlor or metolachlor were apportioned according to the relative abundance of the parent herbicides at this site. Dashed lines represent the U.S. EPA's maximum contaminant level (MCL) for alachlor and atrazine.

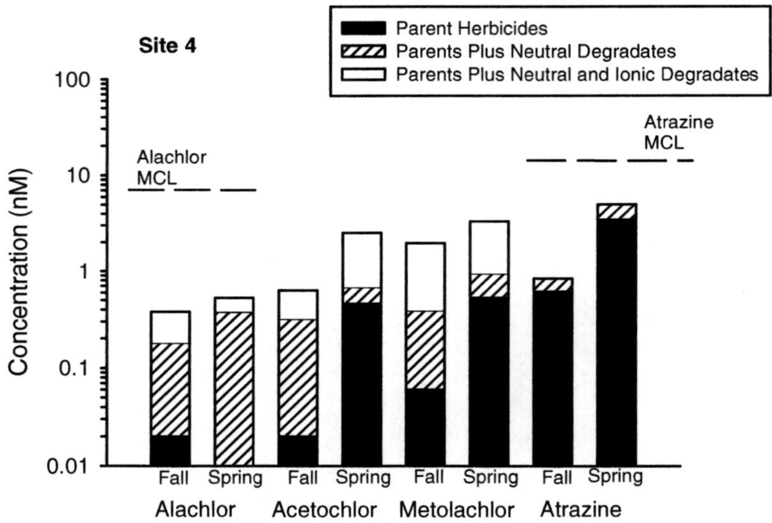

Figure 3.4 Bar chart showing concentrations (on a molar basis) of the parent chloroacetamide and chloro-*s*-triazine herbicides and degradates measured at Site 4 in Fall 2003 and Spring 2004 in the finished water. Those degradates that could originate from either acetochlor or metolachlor were apportioned according to the relative abundance of the parent herbicides at this site. Dashed lines represent the U.S. EPA's maximum contaminant level (MCL) for alachlor and atrazine.

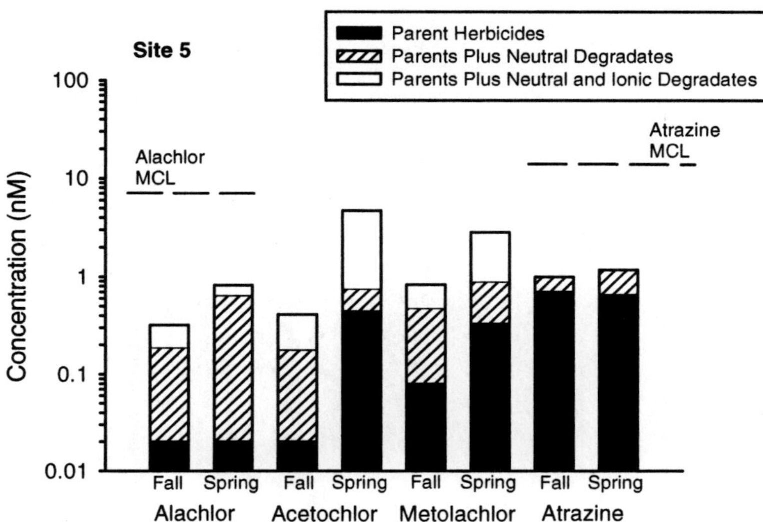

Figure 3.5 Bar chart showing concentrations (on a molar basis) of the parent chloro-acetamide and chloro-*s*-triazine herbicides and degradates measured at Site 5 in Fall 2003 and Spring 2004 in the finished water. Those degradates that could originate from either acetochlor or metolachlor were apportioned according to the relative abundance of the parent herbicides at this site. Dashed lines represent the U.S. EPA's maximum contaminant level (MCL) for alachlor and atrazine.

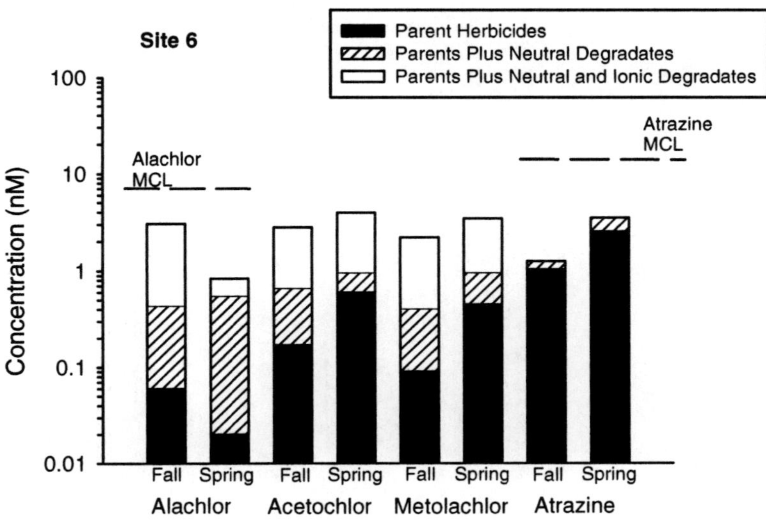

Figure 3.6 Bar chart showing concentrations (on a molar basis) of the parent chloro-acetamide and chloro-*s*-triazine herbicides and degradates measured at Site 6 in Fall 2003 and Spring 2004 in the finished water. Those degradates that could originate from either acetochlor or metolachlor were apportioned according to the relative abundance of the parent herbicides at this site. Dashed lines represent the U.S. EPA's maximum contaminant level (MCL) for alachlor and atrazine.

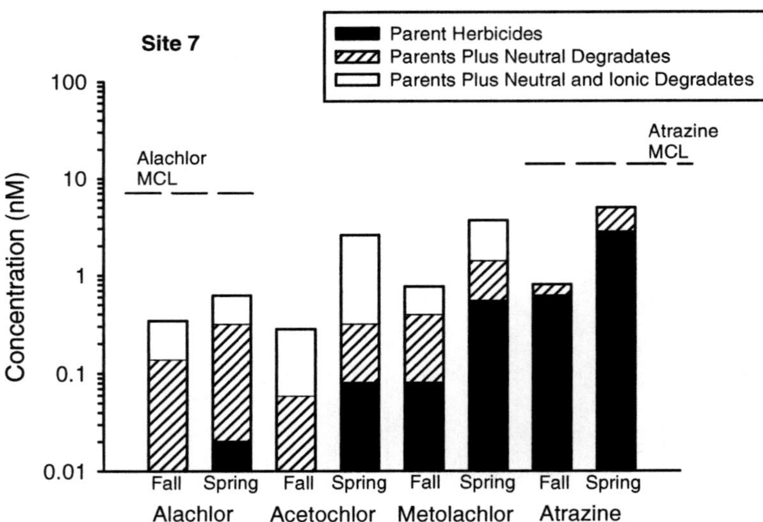

Figure 3.7 Bar chart showing concentrations (on a molar basis) of the parent chloroacetamide and chloro-*s*-triazine herbicides and degradates measured at Site 7 in Fall 2003 and Spring 2004 in the finished water. Those degradates that could originate from either acetochlor or metolachlor were apportioned according to the relative abundance of the parent herbicides at this site. Dashed lines represent the U.S. EPA's maximum contaminant level (MCL) for alachlor and atrazine.

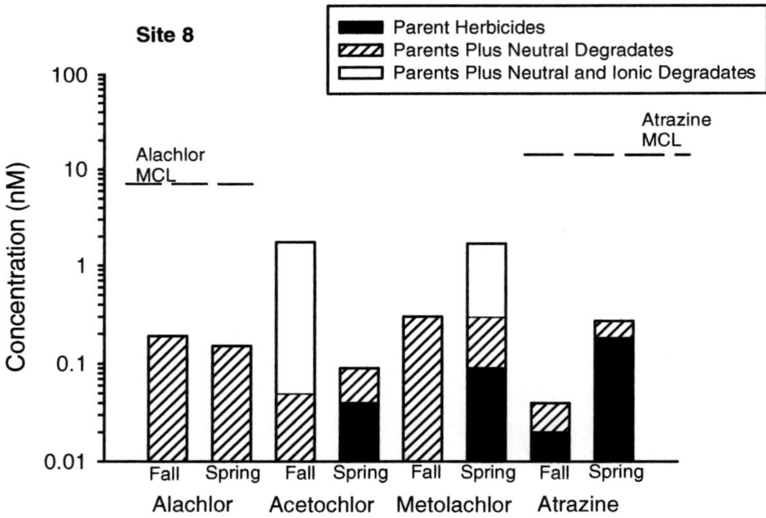

Figure 3.8 Bar chart showing concentrations (on a molar basis) of the parent chloroacetamide and chloro-*s*-triazine herbicides and degradates measured at Site 8 in Fall 2003 and Spring 2004 in the finished water. Those degradates that could originate from either acetochlor or metolachlor were apportioned according to the relative abundance of the parent herbicides at this site. Dashed lines represent the U.S. EPA's maximum contaminant level (MCL) for alachlor and atrazine.

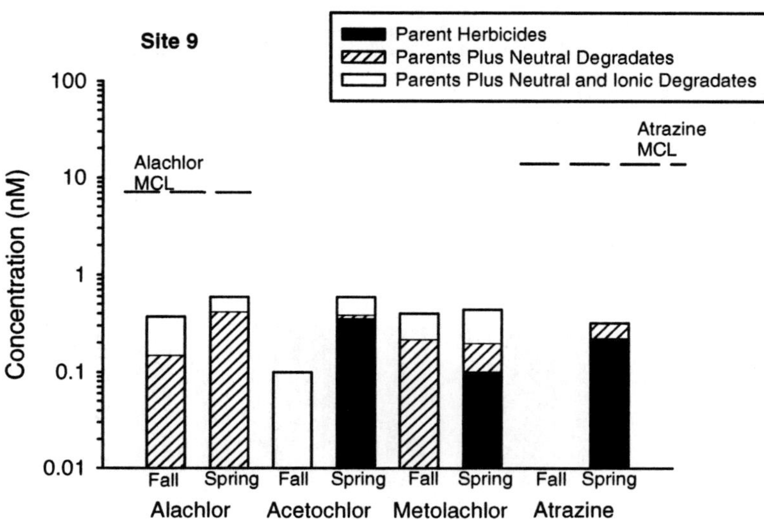

Figure 3.9 Bar chart showing concentrations (on a molar basis) of the parent chloroacetamide and chloro-*s*-triazine herbicides and degradates measured at Site 9 in Fall 2003 and Spring 2004 in the finished water. Those degradates that could originate from either acetochlor or metolachlor were apportioned according to the relative abundance of the parent herbicides at this site. Dashed lines represent the U.S. EPA's maximum contaminant level (MCL) for alachlor and atrazine.

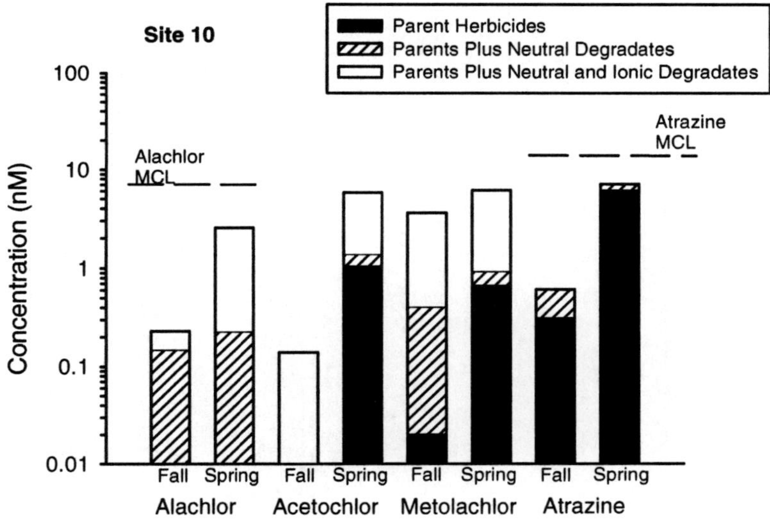

Figure 3.10 Bar chart showing concentrations (on a molar basis) of the parent chloroacetamide and chloro-*s*-triazine herbicides and degradates measured at Site 10 in Fall 2003 and Spring 2004 in the finished water. Those degradates that could originate from either acetochlor or metolachlor were apportioned according to the relative abundance of the parent herbicides at this site. Dashed lines represent the U.S. EPA's maximum contaminant level (MCL) for alachlor and atrazine.

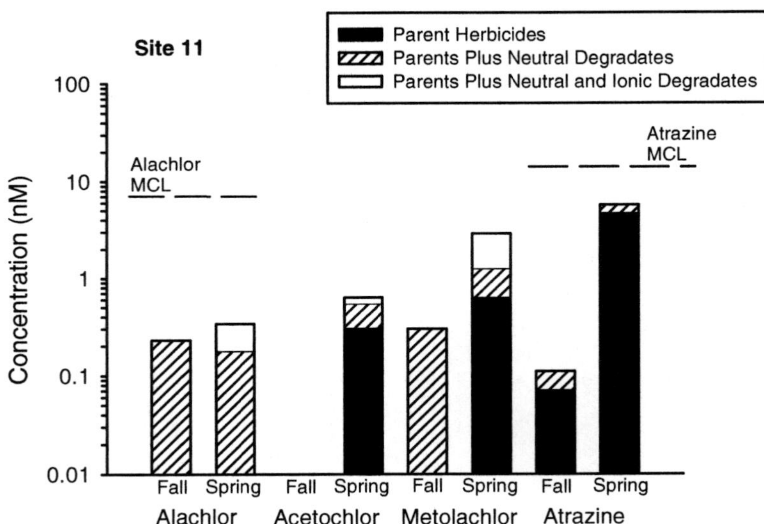

Figure 3.11 Bar chart showing concentrations (on a molar basis) of the parent chloroacetamide and chloro-*s*-triazine herbicides and degradates measured at Site 11 in Fall 2003 and Spring 2004 in the finished water. Those degradates that could originate from either acetochlor or metolachlor were apportioned according to the relative abundance of the parent herbicides at this site. Dashed lines represent the U.S. EPA's maximum contaminant level (MCL) for alachlor and atrazine.

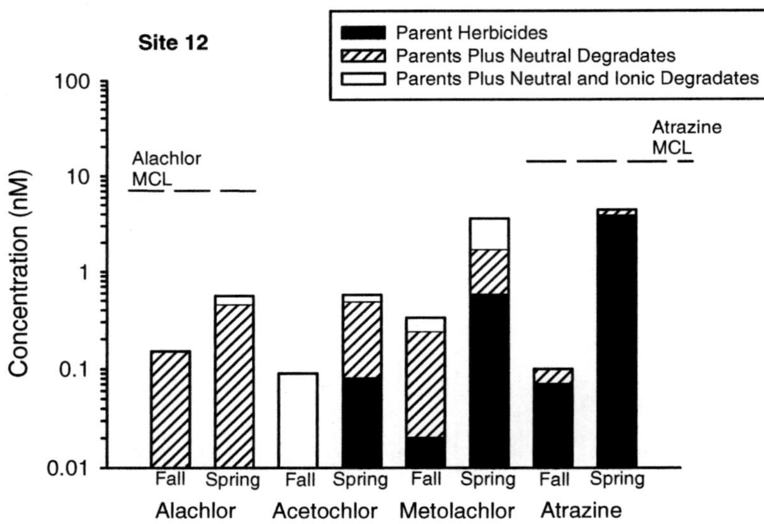

Figure 3.12 Bar chart showing concentrations (on a molar basis) of the parent chloroacetamide and chloro-*s*-triazine herbicides and degradates measured at Site 12 in Fall 2003 and Spring 2004 in the finished water. Those degradates that could originate from either acetochlor or metolachlor were apportioned according to the relative abundance of the parent herbicides at this site. Dashed lines represent the U.S. EPA's maximum contaminant level (MCL) for alachlor and atrazine.

2003); for heuristic purposes we will assume a hypothetical MCL of 2 μg/L. None of the finished water samples would violate such an MCL for acetochlor or metolachlor. If the neutral and ionic degradates were included in an MCL of 2 μg/L for acetochlor or metolachlor, this criterion would not be exceeded in any of the fall or the spring finished water samples.

BENCH SCALE EXPERIMENTS

Coagulation

Coagulation with either alum (30 mg/L) or ferric chloride (20 mg/L) provided little removal (<10%) of the compounds studied (Table 3.5). Removal of the parent chloroacetamides ranged from 4 to 6%. Some of the degradates, such as deschloroacetylmetolachlor propanol, exhibited removal efficiencies as high as 10%. These results indicate that coagulation with alum and ferric chloride is not an effective means of treating the parent chloroacetamide herbicides or their neutral degradates.

Oxidation

Chlorination

Upon aqueous chlorination (6 mg/L applied free chlorine for 6 hrs), 2,6-diethylaniline, deschloroacetylmetolachlor, deschloroacetylmetolachlor propanol, and 2-ethyl-6-methyl aniline, along with dimethenamid and its deschloro degradate, exhibited removal efficiencies of 84 to 100% (Table 3.6). Those degradates of acetochlor, alachlor and metolachlor that lack the acetanilide functional group displayed complete (~100%) removal under the conditions investigated. Those degradates containing the acetanilide functional group, but lacking the *N*-(alkoxy)alkyl side chain, yielded removals of 0-16%. Degradates that still maintained some portion of both the acetanilide and *N*-(alkoxy)alkyl side chain generally displayed removal efficiencies of 1-13%, and most parent herbicides did not undergo detectable removal in the presence of aqueous chlorine. These results suggest that the acetanilide functional group confers a degree of protection against chlorination. Similar studies by other researchers reveal that anilines react readily with aqueous chlorine, while substituted amide compounds display little to no reaction with aqueous chlorine (Katz 1986; Hwang, Larson, and Snoeyink 1990).

One parent chloroacetamide that did react readily (84% removal) with aqueous chlorine was dimethenamid. Its neutral degradate, deschlorodimethenamid, also underwent 96% removal in the presence of aqueous chlorine. These compounds both possess a substituted thienyl ring in place of the alkyl-substituted benzene ring of the other compounds investigated. Product studies undertaken for these two compounds using GC/MS show that the reaction with free chlorine leads to a single observed product in each case, involving addition of a single chlorine onto the thienyl ring (Figure 3.13). GC-NPD analyses of solvent extracts suggest this is the major product; if we assume a similar response for the starting material and the chlorinated product, the molar concentrations of the respective chlorination products are approximately equal to the concentration of the parent compound lost during reaction with aqueous chlorine.

Table 3.5
Removal efficiencies of chloroacetamide herbicides and neutral degradates with alum and ferric chloride during coagulation. Stated uncertainties represent one standard deviation based on triplicate analyses.

Analyte no.	Identity*	Alum† % removal	$FeCl_3$‡ % removal
I	alachlor	4(±1)	4(±1)
II	hydroxyalachlor	6(±1)	5(±1)
III	deschloroalachlor	4(±1)	5(±2)
IV	2-chloro-2'-6'-diethylacetanilide	5(±1)	4(±2)
V	2-hydroxy-2'-6'-diethylacetanilide	5(±2)	7(±1)
VI	2-hydroxy-2'-6'-diethyl-*N*-methylacetanilide	4(±2)	3(±2)
VII	2'-6'-diethylacetanilide	4(±1)	3(±1)
VIII	2,6-diethylaniline	5(±1)	2(±1)
XI	metolachlor	5(±2)	6(±1)
XII	hydroxymetolachlor	8(±3)	9(±2)
XIII	deschlorometolachlor	4(±1)	5(±1)
XIV	metolachlor morpholinone	4(±1)	7(±3)
XV	metolachlor propanol	3(±1)	6(±3)
XVI	deschloroacetylmetolachlor	5(±2)	3(±1)
XVII	deschloroacetylmetolachlor propanol	10(±3)	10(±2)
XX	acetochlor	4(±2)	6(±1)
XXI	hydroxyacetochlor	6(±2)	8(±2)
XXII	deschloroacetochlor	5(±2)	6(±1)
XXV	2-chloro-2'-ethyl-6'-methylacetanilide	3(±1)	5(±1)
XXVI	2-hydroxy-2'-ethyl-6'-methylacetanilide	5(±2)	8(±2)
XXVII	2'-ethyl-6'-methylacetanilide	4(±2)	5(±1)
XXVIII	2-ethyl-6-methylaniline	3(±1)	4(±1)
XXIX	dimethenamid	4(±1)	6(±1)
XXX	deschlorodimethenamid	4(±2)	4(±2)

* Initial concentration = 50 μg/L
† Alum = 30 mg/L
‡ $FeCl_3$ = 20 mg/L

Table 3.6
Removal efficiencies of chloroacetamide herbicides and neutral degradates following application of free chlorine and ozone. Stated uncertainties represent one standard deviation based on triplicate analyses.

Analyte no.	Identity*	Chlorination† % removal	Ozonation‡ % removal
I	alachlor	-1(±2)	63(±1)
II	hydroxyalachlor	7(±2)	70(±1)
III	deschloroalachlor	2(±1)	66(±2)
IV	2-chloro-2'-6'-diethylacetanilide	0(±1)	75(±2)
V	2-hydroxy-2'-6'-diethylacetanilide	13(±4)	77(±1)
VI	2-hydroxy-2'-6'-diethyl-*N*-methylacetanilide	16(±3)	88(±2)
VII	2'-6'-diethylacetanilide	11(±2)	80(±1)
VIII	2,6-diethylaniline	100	100
XI	metolachlor	-2(±2)	60(±2)
XII	hydroxymetolachlor	8(±4)	67(±1)
XIII	deschlorometolachlor	2(±1)	63(±1)
XIV	metolachlor morpholinone	13(±4)	85(±2)
XV	metolachlor propanol	7(±3)	72(±2)
XVI	deschloroacetylmetolachlor	100	100
XVII	deschloroacetylmetolachlor propanol	100	100
XX	acetochlor	-1(±2)	61(±2)
XXI	hydroxyacetochlor	8(±2)	69(±1)
XXII	deschloroacetochlor	1(±1)	64(±1)
XXV	2-chloro-2'-ethyl-6'-methylacetanilide	0(±1)	74(±2)
XXVI	2-hydroxy-2'-ethyl-6'-methylacetanilide	13(±2)	78(±2)
XXVII	2'-ethyl-6'-methylacetanilide	10(±3)	82(±1)
XXVIII	2-ethyl-6-methylaniline	100	100
XXIX	dimethenamid	84(±2)	100
XXX	deschlorodimethenamid	96(±1)	100

* Initial concentration = 50 μg/L
† Applied free chlorine (aqueous HOCl at pH 7); dose = 6 mg/L; contact time = 6 hours
‡ Applied ozone; dose = 3 mg/L; contact time = 30 min

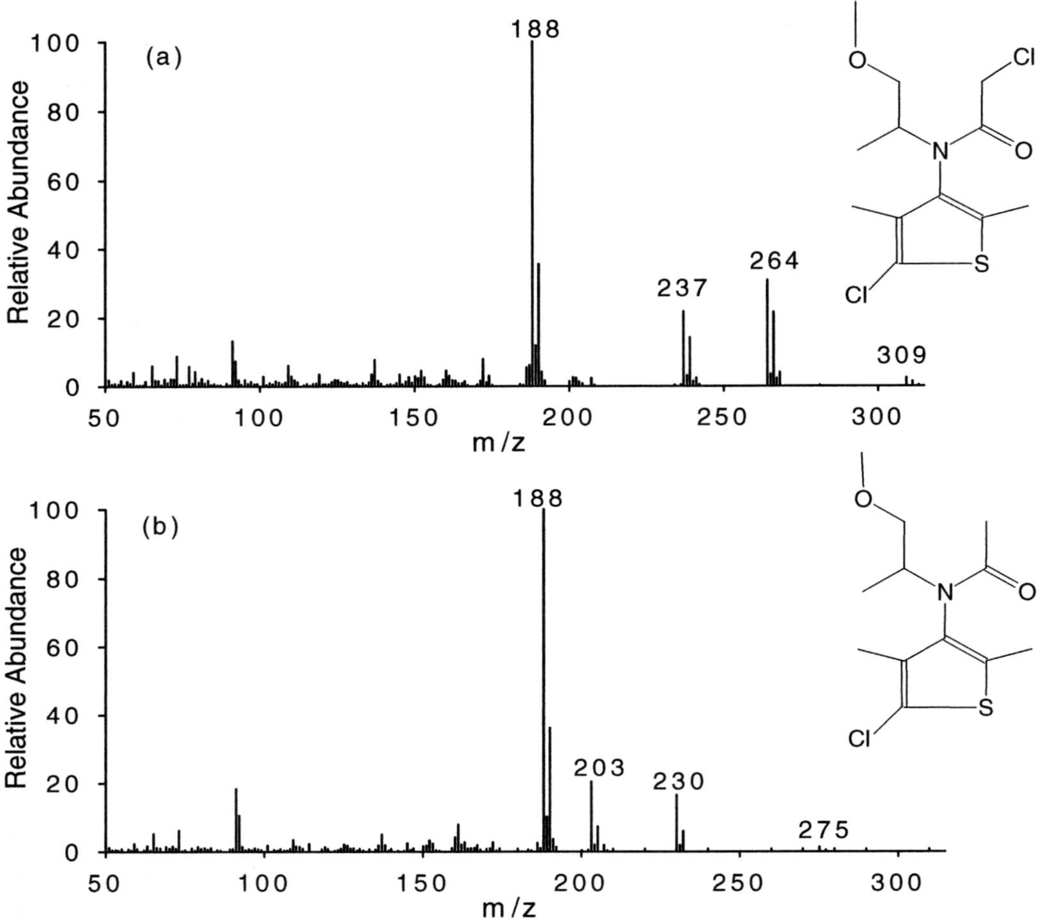

Figure 3.13 Mass spectra of (a) observed dimethenamid product after chlorination and (b) observed deschlorodimethenamid product after chlorination. The products obtained upon chlorination have an additional chlorine on the thienyl ring.

Product studies for the other compounds that reacted with aqueous chlorine were attempted, but we were unable to identify any products using GC/MS. Previous studies of the chlorination of substituted anilines found the products involved chlorine addition to the aniline ring (Hwang, Larson, and Snoeyink 1990). Once formed, it is unlikely that these chlorinated anilines would be further degraded.

Ozonation

All of the compounds reacted with ozone (3 mg/L applied dose for 30 min), exhibiting removals in the range of 60 to 100% (Table 3.6). Those compounds that were more susceptible to reaction with free chlorine showed complete removal on ozonation (100%). Those compounds that did not react with the free chlorine were transformed by contact with ozone (removals of 60 to 88%).

The products formed following ozonation of the test compounds were not identified. Prior research (Somich et al., 1988) into ozonation of alachlor indicates the benzene ring is cleaved, although complete mineralization does not occur. When anilines were reacted with ozone, many degradates have been shown to form, including azobenzenes, azoxybenzenes, and benzidines (Chan and Larson, 1991). These aniline products are *N*-substituted polyaromatic compounds that are likely to be more mutagenic than their parent compounds (Chan and Larson, 1991).

The ozonation conditions tested resulted in higher removal efficiencies for most of the compounds studied relative to removal during chlorination (Table 3.6). Complete mineralization is unlikely to occur under ozonation and chlorination conditions encountered at most water treatment plants. Consequently, these oxidation treatments may not reduce the human health risk associated with the parent herbicides or their neutral degradates.

Activated Carbon

Adsorption data for each compound onto PAC were fit to a Freundlich isotherm. The Freundlich isotherm is an empirical correlation

$$q_e = KC_e^{1/n} \tag{3.1}$$

where q_e = sorption capacity (mg adsorbate per g carbon); C_e = equilibrium solution concentration (mg/L); and K and n are constants. K is primarily related to the capacity of the adsorbent and n is related to the strength of adsorption (Snoeyink, 1990). Example plots of q_e versus C_e can be found elsewhere (Hladik, Roberts and Bouwer, 2005). Such data (when transformed to log-log plots) were used to determine $1/n$ (slope) and K (determined from q_e value where $C_e = 1$).

Values of K and $1/n$ for each compound are listed in Table 3.7. The degradates and parents vary considerably in their affinity for PAC; the parents have a higher adsorption affinity than all of their degradates, with the exception of the morpholinone derivative of metolachlor. When the log K values are compared with the estimated log K_{ow} values, a reasonable correlation ($R^2 = 0.87$) results (Figure 3.14). Those compounds with lower estimated K_{ow} values tend to have lower adsorption affinities for the PAC under test conditions. These lower adsorption affinities for the neutral degradates imply less removal of these compounds is likely than for the parent herbicide at a given PAC loading. Similar results were observed by Gustafson et al. (2003) with the ESA and OA degradates, which are comparatively more water soluble than the parent herbicides, and are less readily removed by sorption onto PAC.

Adsorption by PAC was also explored with atrazine as a sorbate (Table 3.7) as most drinking water systems optimize their PAC addition to achieve a target level of atrazine. For example, if atrazine is assumed to have an influent concentration of 5 μg/L and an effluent concentration of 1 μg/L is desired, our results would yield a recommended PAC dosage of 0.8 mg/L. While the parent chloroacetamides have higher K values than atrazine, resulting in predicted removal efficiencies >80% under these conditions, many of the neutral chloroacetamide degradates have lower K values than atrazine. Given the same influent and desired effluent concentrations, 2-chloro-2'-ethyl-6'-methylacetanilide would require 0.9 mg/L of PAC and 2-hydroxy-2'-ethyl-6'-methyl acetanilide would require 1.5 mg/L of PAC. These values are 1.1 times and 1.9 times the PAC dose needed for comparable control of atrazine.

Table 3.7
Freundlich parameters for adsorption of chloroacetamide herbicides and their neutral degradates onto PAC. Stated uncertainties in K and 1/n indicate 95% confidence intervals.

Analyte no.	Identity*	K†	K† range	$1/n$	$1/n$ range	K_{ow}‡
I	alachlor	266	169-419	0.48	0.41-0.55	1500
II	hydroxyalachlor	127	92-176	0.43	0.37-0.49	560
III	deschloroalachlor	131	118-145	0.46	0.43-0.49	1200
IV	2-chloro-2'-6'-diethylacetanilide	117	117-129	0.44	0.42-0.46	680
V	2-hydroxy-2'-6'-diethylacetanilide	49	39-60	0.38	0.33-0.43	35
VI	2-hydroxy-2'-6'-diethyl-N-methylacetanilide	105	81-135	0.43	0.38-0.48	690
VII	2'-6'-diethylacetanilide	63	48-83	0.41	0.35-0.47	170
VIII	2,6-diethylaniline	121	104-141	0.42	0.39-0.45	740
XI	metolachlor	304	213-433	0.50	0.44-0.56	1800
XII	hydroxymetolachlor	139	104-186	0.44	0.39-0.49	650
XIII	deschlorometolachlor	156	133-182	0.50	0.47-0.53	1400
XIV	metolachlor morpholinone	341	235-493	0.52	0.46-0.58	3800
XV	metolachlor propanol	121	97-151	0.44	0.39-0.49	430
XVI	deschloroacetylmetolachlor	106	80-140	0.42	0.36-0.48	170
XVII	deschloroacetylmetolachlor propanol	53	41-68	0.38	0.32-0.44	39
XX	acetochlor	282	181-439	0.48	0.41-0.55	1500
XXI	hydroxyacetochlor	134	99-181	0.44	0.38-0.50	560
XXII	deschloroacetochlor	123	96-159	0.46	0.41-0.51	1200
XXV	2-chloro-2'-ethyl-6'-methylacetanilide	85	70-103	0.43	0.39-0.47	200
XXVI	2-hydroxy-2'-ethyl-6'-methylacetanilide	30	25-36	0.35	0.30-0.40	10
XXVII	2'-ethyl-6'-methylacetanilide	48	38-61	0.39	0.34-0.44	49
XXVIII	2-ethyl-6-methylaniline	89	64-124	0.43	0.36-0.50	220
XXIX	dimethenamid	129	88-189	0.43	0.36-0.50	300
XXX	deschlorodimethenamid	89	80-98	0.42	0.40-0.44	220
XXXI	atrazine	131	96-179	0.48	0.42-0.54	250

* Initial concentration = 50 μg/L
† Units of K are $(mg/g)/(mg/L)^{1/n}$
‡ K_{ow} estimated using ClogP

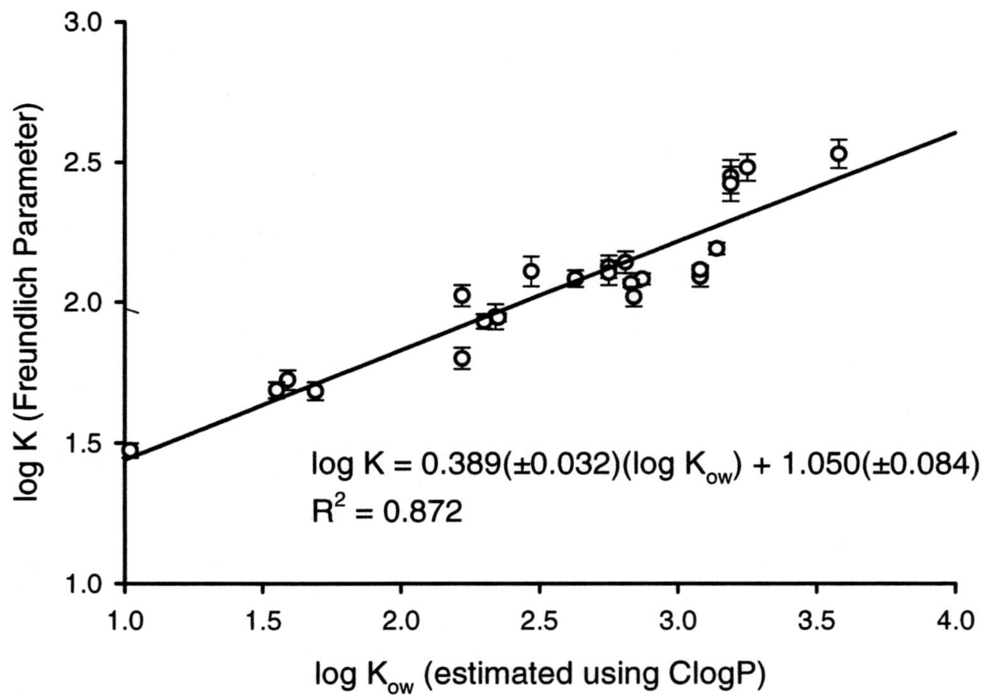

Figure 3.14 Comparison of log K (Freundlich parameter), determined from PAC adsorption studies versus log K_{ow} estimated using ClogP. Error bars represent 95% confidence intervals.

CHAPTER 4
SUMMARY AND CONCLUSIONS

PURPOSE AND APPROACH

This purpose of this project was to develop a method for the analysis of neutral chloroacetamide degradates, to determine the occurrence of the neutral degradates in raw drinking water, and to quantify their removal at the drinking water facilities as well as at the bench scale.

METHOD DEVELOPMENT

For the analysis of chloroacetamide degradates in drinking water where no commercial standards exist, the neutral chloroacetamide degradates were successfully synthesized. Methods were developed for the concentration via solid phase extraction (SPE) and analysis of the neutral chloroacetamide degradates and parent herbicides in drinking water. Concentration of 300-mL water samples was achieved via SPE on Oasis HLB cartridges and analysis was performed using gas chromatography/mass spectrometry (GC/MS). We were able to obtain good to excellent recoveries (69-107%) in deionized water. Good to excellent recoveries (62-107%) were also obtained in a local surface water (obtained from Loch Raven Reservoir, MD) spiked with the target compounds. The local surface water samples also contained preservatives that were used in our drinking water analyses. A repeated analysis of the spiked surface water after 21 days showed results that closely matched those obtained prior to storage. Once the method was optimized, we quantified the method detection limits (MDLs) for neutral degradates and parent herbicides. The MDLs for the hydroxy-substituted compounds are almost two orders of magnitude greater than for the other compounds because of peak tailing as well as extensive fragmentation on the MS. Fortunately we were able to use large volume injections (LVI) of 100 μL aliquots (versus splitless injections of 1 μL aliquots) to decrease the MDLs for almost all of our analytes by nearly two orders of magnitude (MDLs = 4-100 ng/L for splitless injection and 0.06-4 ng/L for LVI). LVI did not prove useful for the two primary anilines (2-ethyl-6-methylaniline and 2,6-diethylaniline), which proved too volatile for analysis via this technique.

Analysis of the ionic chloroacetamide degradates in drinking water was achieved through concentration of 500-mL water samples via SPE on carbon cartridges (ENVI-CARB; Supelco). After extraction, the oxanilic acids (OAs) were methylated (using diazomethane) and were analyzed as their methyl esters using GC/MS, while the ethane sulfonic acids (ESAs) were analyzed using high performance liquid chromatography with diode array detection (HPLC-DAD). Recoveries of 76-96% were obtained by spiking deionized water. The surface water samples yielded recoveries similar to those encountered in deionized water (78-98%). MDLs for the ionic compounds were 7-8 ng/L for the oxanilic acids analyzed using GC/MS and splitless injections (1 μL). For the ESAs the MDLs were 100 ng/L for 100 μL injections onto the HPLC-DAD.

DRINKING WATER TREATMENT FACILITY SAMPLES

The drinking water treatment plants were located in Illinois, Iowa, Indiana, Kentucky and Missouri. Six of the plants relied on surface water as sources, three used groundwater sources, two used a mixture of surface water and groundwater and one facility used surface water that was blended with river bank filtered water. A variety of treatment processes were employed at these plants, such as different coagulants, oxidants and/or the use of activated carbon.

From October to December 2003, we received and processed the first round of samples from these treatment plants. A high frequency of detection was encountered for most of the target compounds (parent herbicides and degradates). All of the parent herbicides and 25 of the 28 degradates under study were detected in at least one sample. Median concentrations were calculated for all of the sites under study in the raw and treated water. For the parent herbicides, atrazine had the highest median concentration in the raw and finished water (~ 100 ng/L) followed by metolachlor (~10 ng/L). The other parent herbicides (cyanazine, simazine, acetochlor, alachlor and dimethenamid) had median concentrations on the same order of magnitude as metolachlor. Among the degradates, the ionic compounds (oxanilic acids and ethane sulfonic acids) had the highest median concentrations (100-1000 ng/L); these were at least one order of magnitude greater than the median concentration of the chloroacetamide parent herbicides. The neutral chloroacetamide and triazine degradates detected had median concentrations near the parent compound (~10 ng/L). Even though these drinking water facilities employed a variety of treatment processes, removals were less than 50% for the parent herbicides or their degradates.

These herbicides are applied to agricultural fields in the spring, and peak concentrations are typically encountered at the end of May into June. From our fall sampling period and analyses, these herbicides and their degradates can still be detected well past this peak runoff period when parent herbicide concentrations in drinking water are considered to be of lesser concern. None of the concentrations encountered for the individual compounds exceed current drinking water MCLs (2 and 3 μg/L for alachlor and atrazine, respectively). Significant compound removal was not achieved with the drinking water treatment processes employed at the facilities studied as they are not optimized for removal of traces of herbicides, especially during the fall and winter.

During June 2004, we received and processed the second round of samples from these treatment plants. A high frequency of detection was encountered for most of the target compounds (parent herbicides and degradates). All of the parent herbicides and 27 of the 28 degradates under study were detected in at least one sample. Median concentrations were calculated for all of the sites under study in the raw and treated water. For the parent herbicides, atrazine had the highest median concentration in the raw and finished water (~2 μg/L) followed by metolachlor and acetochlor (~200-300 ng/L). The other parent herbicides (cyanazine, simazine, alachlor and dimethenamid) had median concentrations of ~10-30 ng/L. Among all the degradates, the ionic chloroacetamide degradates (oxanilic acids and ethane sulfonic acids) had the highest median concentrations (~200-900 ng/L); overall these concentrations were greater than those of the parent chloroacetamide herbicides. The neutral chloroacetamide and triazine degradates detected had concentrations that tended to be lower than those of the parent compounds. Median concentrations were ~10-100 ng/L for the neutral chloroacetamide degradates and ~100-300 ng/L for the atrazine degradates.

Significant average removals of ~40% were observed in the spring for all compounds in comparing the raw and finished drinking water samples in those facilities that employed activated carbon (powdered, granular or a combination of the two). The facility that included ozonation experienced removal of the parent herbicides and some degradates, although other degradates increased in concentration after treatment. Those facilities that employed conventional treatment (coagulation/flocculation, filtration and chlorination) experienced little to no removal for either the parent herbicides or degradates.

A comparison of the results for the June 2004 samples with those obtained in the fall of 2003 reveals that the parent herbicide concentrations are present at much higher concentrations in the spring. This is expected, as these herbicides are applied in early spring, and June is typically when herbicide runoff to nearby waters reaches a maximum. The degradates (neutral and ionic) were also detected in higher concentrations in the June 2004 sampling than in the fall 2003 sampling, but this change was not as dramatic as for the parent herbicides.

BENCH SCALE TREATMENT STUDIES

Coagulation was performed using alum and ferric chloride. Optimal coagulant doses (based on lowest turbidity after coagulation) were 30 mg/L for alum and 20 mg/L for ferric chloride. Under these conditions, little to no removal of any of the compounds was observed.

Oxidation with chlorine (6 mg/L free chlorine, 6 hrs contact time) led to complete removal of those neutral degradates that did not contain the acetanilide functional group. Those compounds that possessed an acetanilide functional group underwent little to no removal (0-16%). Dimethenamid and its deschloro degradate revealed significant removal (84 and 96%, respectively) during chlorination. While these two compounds contain an acetamide functional group, they have a substituted thienyl ring instead of the substituted benzene ring that the other compounds possess. Product studies using gas chromatography/mass spectrometry (GC/MS) of dimethenamid and its deschloro degradate after chlorination revealed that a single product was formed, in each case involving addition of one chlorine to the thienyl ring.

Ozonation (3 mg/L, 30 min contact time) produced extensive removal of all analytes under investigation (>60%). Those compounds that were amenable to removal by chlorination exhibited complete removal upon ozonation, while those compounds not removed during chlorination revealed significant, albeit incomplete, removal upon ozonation.

Adsorption onto activated carbon was explored using powdered activated carbon (PAC). The water containing the target analyte was dosed with PAC concentrations of 0.5 to 8 mg/L. The data were fit using a Freundlich isotherm. While all of the compounds were amenable to adsorption onto PAC, all of the neutral chloroacetamide degradates (with one sole exception) revealed much lower adsorption capacities (given by the logarithm of the Freundlich coefficient K) than the parent herbicides. Log K_{ow} values for each compound were estimated using *ClogP*. A correlation was seen between estimated log K_{ow} values and log K values. Those degradates with lower K_{ow} values are more water soluble than the parent herbicides, and therefore will require more activated carbon than the parent compounds if comparable removals are required. Another isotherm study was performed using atrazine; this is the compound most of the drinking water facilities we studied target for removal. It was found that the parent chloroacetamide herbicides had higher adsorption capacities than atrazine, and they would therefore be removed more extensively than atrazine if PAC is employed to control concentrations of atrazine during "spring flush" periods. About half of the neutral chloroacetamide degradates investigated had higher adsorption capacities than atrazine, and these would also be expected to undergo substantial removal at PAC dosages targeted for atrazine. However, the other half of the neutral chloroacetamide degradates had lower adsorption capacities than atrazine, and some may require as much as twice the atrazine PAC dosage to achieve similar removal.

RESEARCH NEEDS

1. Additional treatments are needed to examine reaction kinetics and trends over time and the look at the influence of oxidant dose and contact time.
2. Treatment studies should be conducted on the products of the reactions.
3. More comprehensive toxicity tests need to be completed to determine if the neutral chloroacetamide degradates are of human health concern at the concentrations we measure (or at concentration likely to be encountered elsewhere).
4. The toxicity of compounds formed during drinking water treatment needs to be examined.
5. Experiments on the removal/behavior of the compounds during membrane treatment, especially reverse osmosis, are needed.
6. Potential biodegradation of the degradates should be studied. Biodegradability is important for behavior in biofilters and in distribution systems.

CHAPTER 5
RECOMMENDATIONS TO UTILITIES

This study was designed to determine whether neutral chloroacetamide degradates are present at appreciable concentrations in raw and finished drinking water. Our study found that these degradates are present in both the raw and finished drinking waters during spring and fall. Currently alachlor is the only chloroacetamide herbicide investigated that is regulated in drinking water; atrazine (a triazine herbicide) and simazine are also regulated. The other three chloroacetamides, the chloroacetamide degradates and the atrazine degradates are not regulated, therefore from a drinking water utility perspective these compounds are not currently recognized as posing a concern. Our work does show that the unregulated parent chloroacetamide herbicides metolachlor and acetochlor are found very frequently and at relatively high concentrations in the spring. While most of the individual degradates do not appear to reach the concentrations of the parent herbicide, they do make up a significant portion of the total herbicide residues during the fall and spring.

If these degradates were to be regulated in the future, our work shows that ozonation and adsorption onto activated carbon represent effective ways of removing both the parents and degradates, although higher PAC doses may be required than for comparable removal of atrazine. Chlorination can remove some compounds, although, our work showed that other compounds (not detected in prior environmental studies) can be generated. Our work did not explore the products formed upon ozonation . Other treatments, such as nanofiltration, may also be effective if chloroacetamide degradates were to be regulated.

For those water utilities located in regions of heavy herbicide use (depicted in Figures 1.1 and 1.2), they should be monitoring for the herbicides and degradates proactively. These utilities should develop a water quality plan so they can increase knowledge about the risks and improve the water quality over time. This will build public trust and confidence that the drinking water is of high quality and safe to drink.

APPENDIX A
TABLES OF HERBICIDE AND HERBICIDE DEGRADATE CONCENTRATIONS

The following pages provide data pertaining to measured concentrations of the parent chloroacetamide and chloro-*s*-triazine herbicides and their degradates measured in raw and treated drinking water samples obtained in Fall of 2003 and Spring of 2004 from 12 Midwestern U.S. drinking water treatment facilities. Also included are concentrations of parent herbicides and neutral degradates in raw and finished drinking water after 28 days of storage at 4°C. Std = standard deviation with n = 3 for all parent herbicides and neutral degradates; for the oxanilic acid and ethane sulfonic acid degradates, n = 2.

Table A.1
Site 1 measurements of target analytes in raw and treated drinking water and the percentage removal of each compound within the drinking water utility during Fall 2003 and Spring 2004. Concentrations are in ng/L.

Site 1		Fall*					Spring†				
#	Compound	Raw	Std	Treated	Std	% Removal	Raw	Std	Treated	Std	% Removal
I	alachlor	16	1	17	1	-5	11	0.7	1.5	0.4	87
II	hydroxyalachlor	28	2	25	1	9	27	1	27	2	-1
III	deschloroalachlor	Nd		Nd			Nd		Nd		
IV	2-chloro-2'-6'-diethylacetanilide	9.7	2.3	12	1	-19	0.56	0.13	0.59	0.13	-5
V	2-hydroxy-2'-6'-diethylacetanilide	16	2	15	2	7	26	2	17	1	34
VI	2-hydroxy-2'-6'-diethyl-*N*-methylacetanilide	Nd		Nd			<11	0	<11	0	
VII	2'-6'-diethylacetanilide	35	2	34	2	1	2.5	0.5	4.4	1	-80
VIII	2,6-diethylaniline	Nd		Nd			Nd		Nd		
IX	alachlor oxanilic acid	69	7	60	1	13	126	4	84	4	33
X	alachlor ethane sulfonic acid	Nd		Nd			Nd		Nd		
XI	metolachlor	51	3	50	1	2	741	6	278	3	62
XII	hydroxymetolachlor	39	2	36	1	8	50	1	26	2	49
XIII	deschlorometolachlor	6.0	1.0	6.3	0.2	-5	4.0	1	3.0	1	25
XIV	metolachlor morpholinone	Nd		Nd			37	1	27	1	28
XV	metolachlor propanol	Nd		Nd			28	4	11	1	60
XVI	deschloroacetylmetolachlor	Nd		Nd			Nd		Nd		
XVII	deschloroacetylmetolachlor propanol	Nd		Nd			Nd		Nd		
XVIII	metolachlor oxanilic acid	68	8	64	6	5	137	3	94	5	31

(continued)

Table A.1 (Continued)

Site 1		Fall*					Spring†				
#	Compound	Raw	Std	Treated	Std	% Removal	Raw	Std	Treated	Std	% Removal
XIX	metolachlor ethane sulfonic acid	444	45	420	3	5	829	3	606	1	27
XX	acetochlor	Nd		Nd			390	6	167	2	57
XXI	hydroxyacetochlor	33	3	29	1	11	87	4	55	4	36
XXII	deschloroacetochlor	Nd		Nd			7.4	0.5	3.0	1	60
XXIII	acetochlor oxanilic acid	67	9	56	1	17	326	11	231	1	29
XXIV	acetochlor ethane sulfonic acid	Nd		Nd			Nd		Nd		
XXV	2-chloro-2'-ethyl-6'-methylacetanilide	167	5	163	11	2	6.2	0.5	1.6	0.4	74
XXVI	2-hydroxy-2'-ethyl-6'-methylacetanilide	33	4	35	2	-6	32	1	17	1	46
XXVII	2'-ethyl-6'-methylacetanilide	44	2	42	5	4	6.1	0.3	12	1	-89
XXVIII	2-ethyl-6-methylaniline	<25		<25			Nd		Nd		
XXIX	dimethenamid	8.2	1.2	7.7	0.4	6	14	1	8.7	0.6	39
XXX	deschlorodimethenamid	9.8	0.6	8.7	1.0	11	Nd		Nd		
XXXI	atrazine	102	7	89	7	12	3150	30	1260	10	60
XXXII	desethyl atrazine	14	2	13	0	2	237	30	199	11	16
XXXIII	desisopropyl atrazine	Nd		Nd			90	6.2	62	4	31
XXXIV	simazine	6.6	1.0	5.7	0.2	13	17	0.2	6.8	0.6	60
XXXV	cyanazine	Nd		Nd			9.6	1.3	3.4	4.6	64

NOTE: Water samples from surface water

* Fall 2003 treatment train: lime softener, ferric chloride, CO_2-recarbonation, fluoride, chlorine

† Spring 2004 treatment train: cationic polymer, lime softener, ferric chloride, PAC, CO_2-recarbonation, chlorine, fluoride, phosphate, CO_2, chlorine, GAC capped sand filters, ammonium sulfate, chlorine

Table A.2

Site 2 measurements of target analytes in raw and treated drinking water and the percentage removal of each compound within the drinking water utility during Fall 2003 and Spring 2004. Concentrations are in ng/L.

Site 2		Fall*					Spring†				
#	Compound	Raw	Std	Treated	Std	% Removal	Raw	Std	Treated	Std	% Removal
I	alachlor	8.1	0.6	Nd		>97	4.4	0.1	1.6	0.2	64
II	hydroxyalachlor	Nd		Nd			29	1	25	2	12
III	deschloroalachlor	Nd		Nd			Nd		Nd		
IV	2-chloro-2'-6'-diethylacetanilide	11	1	6.7	1.3	41	0.50	0.06	<0.4		
V	2-hydroxy-2'-6'-diethylacetanilide	11	1	8.8	0.7	22	51	4	29	2	43
VI	2-hydroxy-2'-6'-diethyl-*N*-methylacetanilide	Nd		Nd			Nd		Nd		
VII	2'-6'-diethylacetanilide	32	1	27	1	16	5.4	0.9	2.9	0.4	46
VIII	2,6-diethylaniline	Nd		Nd			Nd		Nd		
IX	alachlor oxanilic acid	21	1	24	1	-18	216	2	136	6	37
X	alachlor ethane sulfonic acid	402	8	361	1	10	460	29	302	3	34
XI	metolachlor	6.6	0.6	7.1	0.6	-7	523	25	260	29	50
XII	hydroxymetolachlor	49	2	18	1	64	35	1	24	2	31
XIII	deschlorometolachlor	<0.55		<0.55			5.9	1	2.8	0.6	53
XIV	metolachlor morpholinone	34	1	Nd		>99	60	3	31	3	50
XV	metolachlor propanol	Nd		Nd			59	3	34	2	43
XVI	deschloroacetylmetolachlor	0.87	0.26	Nd		>67	Nd		Nd		
XVII	deschloroacetylmetolachlor propanol	Nd		Nd			Nd		Nd		
XVIII	metolachlor oxanilic acid	30	1	31	0	-3	164	2	113	3	31

(continued)

Table A.2 (Continued)

Site 2		Fall*					Spring†				
#	Compound	Raw	Std	Treated	Std	% Removal	Raw	Std	Treated	Std	% Removal
XIX	metolachlor ethane sulfonic acid	<276		<276			1260	10	975	2	23
XX	acetochlor	Nd		Nd			990	3	372	19	62
XXI	hydroxyacetochlor	Nd		Nd			110	3	62	5	43
XXII	deschloroacetochlor	Nd		Nd			14	2	4.9	0.6	66
XXIII	acetochlor oxanilic acid	32	1	36	1	-11	638	11	463	12	27
XXIV	acetochlor ethane sulfonic acid	560	39	528	12	6	1080	10	688	36	36
XXV	2-chloro-2'-ethyl-6'-methylacetanilide	22	1	23	2	-3	22	2	2.7	5.8	88
XXVI	2-hydroxy-2'-ethyl-6'-methylacetanilide	4.0	0.8	11	1	-169	54	4	34	4	37
XXVII	2'-ethyl-6'-methylacetanilide	43	1	39	1	9	11	1	6.2	0.5	44
XXVIII	2-ethyl-6-methylaniline	<25		<25			Nd		<25		
XXIX	dimethenamid	Nd		Nd			308	21	5.8	1.4	98
XXX	deschlorodimethenamid	5.2	0.2	4.8	0.3	7	Nd		Nd		
XXXI	atrazine	39	1	39	2	-1	3250	60	1170	38	64
XXXII	desethyl atrazine	11	1	16	0	-39	447	19	125	10	72
XXXIII	desisopropyl atrazine	Nd		6.3		<-93	120	8	27	2	78
XXXIV	simazine	3.0	0.3	3.1	0.2	-5	25	2	12	1	54
XXXV	cyanazine	8.7	0.4	8.3	0.1	5	6.0	3	Nd		>94

NOTE: Water samples from surface water

* Fall 2003 treatment train: potassium permanganate, cationic polymer, poly-aluminum sulfate, anionic polymer, chlorine, GAC, chlorine, ammonia, caustic soda, fluoride, zinc orthophosphate

† Spring 2004 treatment train: pre-chlorine, potassium permanganate, cationic polymer, poly-aluminum sulfate, anionic polymer, chlorine, GAC, chlorine, ammonia, caustic soda, fluoride, zinc orthophosphate

Table A.3
Site 3 measurements of target analytes in raw and treated drinking water and the percentage removal of each compound within the drinking water utility during Fall 2003 and Spring 2004. Concentrations are in ng/L.

Site 3		Fall[*]					Spring[†]				
#	Compound	Raw	Std	Treated	Std	% Removal	Raw	Std	Treated	Std	% Removal
I	alachlor	9.2	0.3	7.9	1.5	13	3.5	0.1	2.0	0.3	41
II	hydroxyalachlor	19	1	Nd		>55	28	2	Nd		
III	deschloroalachlor	Nd		Nd			14	0.58	Nd		>95
IV	2-chloro-2'-6'-diethylacetanilide	16	2	11	1	28	1.6	0.06	1.2	0.2	23
V	2-hydroxy-2'-6'-diethylacetanilide	5.1	2.2	9.2	1.5	-80	104	7	24	1	77
VI	2-hydroxy-2'-6'-diethyl-*N*-methylacetanilide	Nd		Nd			Nd		Nd		
VII	2'-6'-diethylacetanilide	43	2	38	2	13	3.3	0.3	3.0	0.2	9
VIII	2,6-diethylaniline	Nd		Nd			Nd		Nd		
IX	alachlor oxanilic acid	49	1	46	1	6	180	7	117	5	35
X	alachlor ethane sulfonic acid	Nd		Nd			Nd		Nd		
XI	metolachlor	20	1	16	1	20	578	16	311	15	46
XII	hydroxymetolachlor	33	1	25	2	24	42	2	35	1	16
XIII	deschlorometolachlor	3.6	0.5	2.8	0.5	22	4.8	0.2	4.3	0.3	9
XIV	metolachlor morpholinone	Nd		Nd			62	1.6	36	1	43
XV	metolachlor propanol	Nd		Nd			47	3.8	31	1	33
XVI	deschloroacetylmetolachlor	0.89	0.16	2.8	1.4	-211	1.3	0.3	1.4	0.1	-6
XVII	deschloroacetylmetolachlor propanol	Nd		Nd			Nd		Nd		
XVIII	metolachlor oxanilic acid	54	5	51	4	6	154	8	112	3	27

(continued)

Table A.3 (Continued)

Site 3		Fall*					Spring†				
#	Compound	Raw	Std	Treated	Std	% Removal	Raw	Std	Treated	Std	% Removal
XIX	metolachlor ethane sulfonic acid	491	10	387	11	21	1126	20	911	5	19
XX	acetochlor	6.5	0.8	4.5	0.7	31	476	11	269	8	44
XXI	hydroxyacetochlor	39	2	32	2	18	119	3	59	3	51
XXII	deschloroacetochlor	27	2	20	4	25	14	2	6.3	0.4	56
XXIII	acetochlor oxanilic acid	55	6	52	3	6	339	18	235	6	31
XXIV	acetochlor ethane sulfonic acid	Nd		Nd			576	14	391	1	32
XXV	2-chloro-2'-ethyl-6'-methylacetanilide	43	1	37	2	13	18	1	5.2	0.7	71
XXVI	2-hydroxy-2'-ethyl-6'-methylacetanilide	2.7	0.8	9.0	1.1	-231	41	2	24	1	41
XXVII	2'-ethyl-6'-methylacetanilide	57	2	57	3	0	9.6	0.2	9.5	0..3	1
XXVIII	2-ethyl-6-methylaniline	<25		<25			Nd		Nd		
XXIX	dimethenamid	6.8	0.8	Nd		>96	124	3	46	7	63
XXX	deschlorodimethenamid	14	2	25	2	-78	0.32	0.04	Nd		>6
XXXI	atrazine	133	1	106	2	20	4250	40	1860	30	56
XXXII	desethyl atrazine	44	3	44	3	0	308	10	263	19	15
XXXIII	desisopropyl atrazine	10	1	11	2	-10	104	5	63	3	40
XXXIV	simazine	12	1	12	1	0	35	1	20	1	41
XXXV	cyanazine	Nd		Nd			Nd		Nd		

NOTE: Water samples from surface water

* Fall 2003 treatment train: PAC, chlorine, ammonia, Clarion 410p coagulant, filter (with GAC), zinc ortho, fluoride, chlorine

† Spring 2004 treatment train: PAC, chlorine, ammonia, alum/poly blend, sand and GAC filter, zinc ortho, fluoride, chlorine, ammonia

Table A.4
Site 4 measurements of target analytes in raw and treated drinking water and the percentage removal of each compound within the drinking water utility during Fall 2003 and Spring 2004. Concentrations are in ng/L.

Site 4		Fall*					Spring†				
#	Compound	Raw	Std	Treated	Std	% Removal	Raw	Std	Treated	Std	% Removal
I	alachlor	6.8	1.3	6.6	0.9	3	2.3	0.5	1.7	0.7	26
II	hydroxyalachlor	Nd		Nd			Nd		Nd		
III	deschloroalachlor	Nd		Nd			Nd		Nd		
IV	2-chloro-2'-6'-diethylacetanilide	5.5	0.8	4.7	0.2	15	1.5	0.1	1.7	0.1	-18
V	2-hydroxy-2'-6'-diethylacetanilide	Nd		Nd			68	3	61	5	12
VI	2-hydroxy-2'-6'-diethyl-*N*-methylacetanilide	Nd		Nd			Nd		Nd		
VII	2'-6'-diethylacetanilide	24	1	26	1	-9	7.3	0.4	8.9	0.6	-22
VIII	2,6-diethylaniline	Nd		Nd			Nd		Nd		
IX	alachlor oxanilic acid	52	2	54	1	-4	43	2	41	0	5
X	alachlor ethane sulfonic acid	Nd		Nd			Nd		Nd		
XI	metolachlor	16	1	16	1	3	165	3	149	2	9
XII	hydroxymetolachlor	26	3	24	3	7	29	0	30	3	-4
XIII	deschlorometolachlor	3.0	1	2.9	1	3	4.9	0.6	4.8	0.2	3
XIV	metolachlor morpholinone	Nd		Nd			26	1	21	2	19
XV	metolachlor propanol	Nd		Nd			34	1	28	2	18
XVI	deschloroacetylmetolachlor	1.2	1.3	2.5	0.3	-116	2.4	0.8	4.1	0.3	-69
XVII	deschloroacetylmetolachlor propanol	Nd		Nd			Nd		Nd		
XVIII	metolachlor oxanilic acid	74	0	74	0	1	243	10	148	25	39

(continued)

Table A.4 (Continued)

Site 4		Fall*					Spring†				
#	Compound	Raw	Std	Treated	Std	% Removal	Raw	Std	Treated	Std	% Removal
XIX	metolachlor ethane sulfonic acid	438	27	420	1	4	902	17	594	3	34
XX	acetochlor	6.9	0.4	6.6	1.0	6	132	7	122	5	7
XXI	hydroxyacetochlor	40	3	38	3	4	46	1	37	1	20
XXII	deschloroacetochlor	13	1	14	1	-8	4.0	0.9	4.4	0.1	-9
XXIII	acetochlor oxanilic acid	86	2	82	2	4	288	1	204	39	29
XXIV	acetochlor ethane sulfonic acid	Nd		Nd			642	6	327	16	49
XXV	2-chloro-2'-ethyl-6'-methylacetanilide	16	0	22	1	-38	0.5	0.4	0.9	0.6	-95
XXVI	2-hydroxy-2'-ethyl-6'-methylacetanilide	Nd		Nd			17	1	16	1	4
XXVII	2'-ethyl-6'-methylacetanilide	36	2	35	1	0	9.0	1	8.2	0.7	8
XXVIII	2-ethyl-6-methylaniline	<25		<25			Nd		Nd		
XXIX	dimethenamid	3.4	1.8	2.7	1.7	21	7.1	0.7	Nd		>96
XXX	deschlorodimethenamid	4.9	2.3	2.0	0.6	59	1.4	0.1	1.1	0.3	18
XXXI	atrazine	135	2	131	2	3	785	17	740	13	6
XXXII	desethyl atrazine	33	2	31	1	8	252	33	228	5	9
XXXIII	desisopropyl atrazine	11	1	10	1	4	63	3	57	2	10
XXXIV	simazine	10	1	9.8	0.6	6	15	1	13	1	12
XXXV	cyanazine	Nd		Nd			Nd		Nd		

NOTE: Water samples from surface water

* Fall 2003 treatment train: alum, cationic polymer, chlorine, ammonia

† Spring 2004 treatment train: alum, cationic polymer, chlorine, ammonia

Table A.5

Site 5 measurements of target analytes in raw and treated drinking water and the percentage removal of each compound within the drinking water utility during Fall 2003 and Spring 2004. Concentrations are in ng/L.

Site 5		Fall*					Spring†				
#	Compound	Raw	Std	Treated	Std	% Removal	Raw	Std	Treated	Std	% Removal
I	alachlor	5.7	0.2	5.0	0.2	13	6.0	0.8	5.4	0.4	9
II	hydroxyalachlor	Nd		Nd			43	2	34	4	20
III	deschloroalachlor	Nd		Nd			Nd		Nd		
IV	2-chloro-2'-6'-diethylacetanilide	11	0.3	8.5	0.6	21	1.7	0.1	1.7	0.1	-2
V	2-hydroxy-2'-6'-diethylacetanilide	Nd		Nd			82	3	85	4	-3
VI	2-hydroxy-2'-6'-diethyl-*N*-methylacetanilide	Nd		Nd			<11		<11		
VII	2'-6'-diethylacetanilide	35	1.3	25	1	29	13	2	8.2	1.0	34
VIII	2,6-diethylaniline	Nd		Nd			Nd		Nd		
IX	alachlor oxanilic acid	40	5	34	1	15	140	8	46	3	68
X	alachlor ethane sulfonic acid	Nd		Nd			Nd		Nd		
XI	metolachlor	30	1	24	1	21	891	17	94	2	89
XII	hydroxymetolachlor	29	1	28	2	6	156	3	30	2	81
XIII	deschlorometolachlor	1.0	0.4	2.5	0.5	-137	9.9	0.6	8.5	0.1	14
XIV	metolachlor morpholinone	Nd		Nd			63	1	27	2.1	58
XV	metolachlor propanol	Nd		Nd			112	10	55	3.	51
XVI	deschloroacetylmetolachlor	2.4	0.9	1.6	0.7	33	8.7	1.7	7.1	0.6	19
XVII	deschloroacetylmetolachlor propanol	Nd		Nd			Nd		Nd		
XVIII	metolachlor oxanilic acid	104	10	98	2	6	687	80	168	24	76

(continued)

Table A.5 (Continued)

Site 5 #	Compound	Fall* Raw	Std	Treated	Std	% Removal	Spring† Raw	Std	Treated	Std	% Removal
XIX	metolachlor ethane sulfonic acid	353		<276			772	30	437	16	43
XX	acetochlor	4.7	0.6	4.5	0.7	6	313	12	117	4	63
XXI	hydroxyacetochlor	22	1	19	1	15	198	2	51	3	74
XXII	deschloroacetochlor	Nd		8	0	<-98	14	1	5.8	0.2	59
XXIII	acetochlor oxanilic acid	58	0	62	1	-7	1170	110	551	11	53
XXIV	acetochlor ethane sulfonic acid	Nd		Nd			1050	10	578	10	45
XXV	2-chloro-2'-ethyl-6'-methylacetanilide	28	2	26	1	7	10	2	6.0	0.4	41
XXVI	2-hydroxy-2'-ethyl-6'-methylacetanilide	Nd		Nd			105	6	15	1	86
XXVII	2'-ethyl-6'-methylacetanilide	53	2	37	2	31	11	1	11	1	1
XXVIII	2-ethyl-6-methylaniline	<25		<25			Nd		Nd		
XXIX	dimethenamid	2.5	0.3	2.2	0.4	10	4.0	0.4	3.3	0.3	17
XXX	deschlorodimethenamid	4.7	1.0	2.5	0.8	48	1.4	0.2	1.3	0.1	8
XXXI	atrazine	201	8	151	3	25	1760	20	139	5	92
XXXII	desethyl atrazine	32	2	46	2	-43	594	1	74	2	88
XXXIII	desisopropyl atrazine	7.4	0.2	9.9	1.6	-34	180	9	24	1	87
XXXIV	simazine	8.6	0.7	6.3	0.2	27	69	3	18	1	74
XXXV	cyanazine	8.7	0.5	7.9	0.2	8	Nd		Nd		

NOTE: Water samples from surface water

* Fall 2003 treatment train: polymer, catfloc, polydyne, lime, ferric Chloride, orthophosphate, chlorine, fluoride

† Spring 2004 treatment train: polymer, catfloc, lime, ferric chloride, GAC filters, orthophosphate, chlorine, fluoride

Table A.6

Site 6 measurements of target analytes in raw and treated drinking water and the percentage removal of each compound within the drinking water utility during Fall 2003 and Spring 2004. Concentrations are in ng/L.

Site 6		Fall*					Spring†				
#	Compound	Raw	Std	Treated	Std	% Removal	Raw	Std	Treated	Std	% Removal
I	alachlor	17	1	17	1	4	6.9	0.7	5.6	0.3	20
II	hydroxyalachlor	46	3	44	2	3	36	4	30	3	16
III	deschloroalachlor	Nd		Nd			Nd		Nd		
IV	2-chloro-2'-6'-diethylacetanilide	6.3	0.5	9.8	1.1	-55	1.8	0.2	0.99	0.13	45
V	2-hydroxy-2'-6'-diethylacetanilide	Nd		Nd			92	4	74	2	19
VI	2-hydroxy-2'-6'-diethyl-*N*-methylacetanilide	Nd		Nd			<11		<11		
VII	2'-6'-diethylacetanilide	26	1	30	1	-14	7.7	0.9	6.3	0.1	19
VIII	2,6-diethylaniline	Nd		Nd			Nd		Nd		
IX	alachlor oxanilic acid	58	2	69	1.0	-18	110	12	75	2	32
X	alachlor ethane sulfonic acid	945	82	743	14	21	Nd		Nd		
XI	metolachlor	27	2	25	1	8	431	7	128	15	70
XII	hydroxymetolachlor	33	2	32	5	5	61	5	35	3	43
XIII	deschlorometolachlor	1.4	0.3	5.0	1	-270	8.6	0.5	6.7	0.4	22
XIV	metolachlor morpholinone	Nd		Nd			55	2	37	2	33
XV	metolachlor propanol	Nd		15	1	<-96	46	2	30	2	34
XVI	deschloroacetylmetolachlor	1.9	0.7	3.0	0.5	-60	7.6	0.6	4.9	0.3	35
XVII	deschloroacetylmetolachlor propanol	Nd		Nd			Nd		Nd		
XVIII	metolachlor oxanilic acid	85	1	137	5	-61	288	3	215	1	25

(continued)

Table A.6 (Continued)

Site 6		Fall*					Spring†				
#	Compound	Raw	Std	Treated	Std	% Removal	Raw	Std	Treated	Std	% Removal
XIX	metolachlor ethane sulfonic acid	531	27	429	29	19	888	29	562	25	37
XX	acetochlor	56	1	45	1	20	356	5	161	1	55
XXI	hydroxyacetochlor	60	2	55	2	7	96	4	64	25	33
XXII	deschloroacetochlor	14	1	13	1	9	10	1	5.6	1.1	47
XXIII	acetochlor oxanilic acid	68	1	81	4	-19	725	4	477	26	34
XXIV	acetochlor ethane sulfonic acid	645	61	574	20	11	634	4	368	3	42
XXV	2-chloro-2'-ethyl-6'-methylacetanilide	20	1	23	3	-17	9.1	1.6	3.9	2.1	58
XXVI	2-hydroxy-2'-ethyl-6'-methylacetanilide	3.5	2.3	Nd		>37	50	2	20	1	60
XXVII	2'-ethyl-6'-methylacetanilide	37	2	40	2	-10	7.2	1.0	6.4	0.2	11
XXVIII	2-ethyl-6-methylaniline	<25		<25			Nd		Nd		
XXIX	dimethenamid	4.6	0.5	4.9	1	-5	14	1	2.8	2.1	80
XXX	deschlorodimethenamid	7.7	1.9	5.8	1	24	1.5	0.2	1.3	0.1	15
XXXI	atrazine	266	5	225	4	15	1950	40	542	12	72
XXXII	desethyl atrazine	18	1	29	1	-61	281	16	136	11	52
XXXIII	desisopropyl atrazine	9.9	0.8	11	1	-8	95	4	42	1	56
XXXIV	simazine	22	3	19	1	13	29	3	22	1	27
XXXV	cyanazine	12	1	11	1	7	Nd		Nd		

NOTE: Water samples from surface water/groundwater (78:22 in fall; 75:25 in spring)

* Fall 2003 treatment train: coagulant, filtration with GAC, chlorine

† Spring 2004 treatment train: polyaluminum hydroxychloride/polyquatenaryamine blend coagulant, PAC, potassium permanganate, copper sulfate, chlorine, filtration with GAC, chlorine

Table A.7

Site 7 measurements of target analytes in raw and treated drinking water and the percentage removal of each compound within the drinking water utility during Fall 2003 and Spring 2004. Concentrations are in ng/L.

Site 7		Fall*					Spring†				
#	Compound	Raw	Std	Treated	Std	% Removal	Raw	Std	Treated	Std	% Removal
I	alachlor	1.3	0.2	1.2	0.2	2	5.7	0.8	4.5	0.5	21
II	hydroxyalachlor	11	3	5.2	0.5	52	41	1	31	1	25
III	deschloroalachlor	Nd		Nd			3.9	0.3	0.73	0.14	81
IV	2-chloro-2'-6'-diethylacetanilide	6.3	0.2	1.7	1	73	3.7	0.2	3.4	0.2	8
V	2-hydroxy-2'-6'-diethylacetanilide	Nd		Nd			32	3	23	1	27
VI	2-hydroxy-2'-6'-diethyl-*N*-methylacetanilide	Nd		Nd			<11		<11		
VII	2'-6'-diethylacetanilide	24	1	22	1	9	9.9	0.7	8.6	0.1	13
VIII	2,6-diethylaniline	Nd		Nd			Nd		Nd		
IX	alachlor oxanilic acid	62	4	53	2	15	120	11	80	6	33
X	alachlor ethane sulfonic acid	Nd		Nd			Nd		Nd		
XI	metolachlor	26	1	24	1	9	887	7	154	2	83
XII	hydroxymetolachlor	14	1	11	1	25	217	2	61	2	72
XIII	deschlorometolachlor	6.0	1	5.2	0.2	13	15	2	6.2	0.7	59
XIV	metolachlor morpholinone	12	1	8.8	0.4	24	62	5	33	3	47
XV	metolachlor propanol	Nd		Nd			208	9	73	3	65
XVI	deschloroacetylmetolachlor	5.1	0.4	5.6	0.8	-10	17	2	5.3	0.7	69
XVII	deschloroacetylmetolachlor propanol	Nd		Nd			<2.5		<2.5		
XVIII	metolachlor oxanilic acid	138	5	99	4	28	205	2	113	3	45

(continued)

Table A.7 (Continued)

Site 7		Fall*					Spring†				
#	Compound	Raw	Std	Treated	Std	% Removal	Raw	Std	Treated	Std	% Removal
XIX	metolachlor ethane sulfonic acid	324	12	Nd		>1	942	11	599	18	36
XX	acetochlor	0.93	0.15	0.51	0.25	46	180	6	20	3	89
XXI	hydroxyacetochlor	17	1	13	2	23	167	5	48	2	71
XXII	deschloroacetochlor	Nd		Nd			15	1	6.3	0.7	58
XXIII	acetochlor oxanilic acid	76	8	59	2	23	194	18	129	1	33
XXIV	acetochlor ethane sulfonic acid	Nd		Nd			1010	29	557	32	45
XXV	2-chloro-2'-ethyl-6'-methylacetanilide	11	1	8.8	0.9	22	33	1	6.8	0.2	80
XXVI	2-hydroxy-2'-ethyl-6'-methylacetanilide	Nd		Nd			87	4	21	1	75
XXVII	2'-ethyl-6'-methylacetanilide	31	1	28	1	8	15	1	14	1	11
XXVIII	2-ethyl-6-methylaniline	<25		<25			Nd		Nd		
XXIX	dimethenamid	1.4	0.2	1.5	0.4	-8	12	1	1.4	0.2	89
XXX	deschlorodimethenamid	2.1	0.6	2.2	0.5	-3	1.8	0.2	1.1	0.2	37
XXXI	atrazine	186	4	131	1	29	2240	10	590	5	74
XXXII	desethyl atrazine	56	1	26	2	53	477	10	318	6	33
XXXIII	desisopropyl atrazine	14	1	9.0	0.9	37	165	7	75	5	54
XXXIV	simazine	49	1	32	1	34	138	2	69	1	50
XXXV	cyanazine	11	1	9.3	1.3	13	19	1	11	1	42

NOTE: Water samples from surface water/groundwater (85:25)

* Fall 2003 treatment train: potassium permanganate, caustic soda, chlorine, coagulant, filtration with GAC, chlorine, ammonia, phosphate, fluoride, caustic soda

† Spring 2004 treatment train: PAC, chlorine, coagulant/coagulant aid, filtration with GAC, chlorine, ammonia, phosphate, fluoride, caustic soda

Table A.8

Site 8 measurements of target analytes in raw and treated drinking water and the percentage removal of each compound within the drinking water utility during Fall 2003 and Spring 2004. Concentrations are in ng/L.

Site 8		Fall*					Spring†				
#	Compound	Raw	Std	Treated	Std	% Removal	Raw	Std	Treated	Std	% Removal
I	alachlor	1.7	0.7	1.5	0.4	12	0.46	0.11	Nd		>63
II	hydroxyalachlor	Nd		Nd			Nd		Nd		
III	deschloroalachlor	Nd		Nd			Nd		Nd		
IV	2-chloro-2'-6'-diethylacetanilide	4.9	0.2	6.1	0.2	-25	0.70	0.06	2.0	0.3	-182
V	2-hydroxy-2'-6'-diethylacetanilide	Nd		Nd			17	0	20	1	-18
VI	2-hydroxy-2'-6'-diethyl-*N*-methylacetanilide	Nd		Nd			Nd		Nd		
VII	2'-6'-diethylacetanilide	28	1	28	1	0	5.1	0.4	7.9	0.3	-56
VIII	2,6-diethylaniline	Nd		Nd			Nd		Nd		
IX	alachlor oxanilic acid	Nd		Nd			Nd		Nd		
X	alachlor ethane sulfonic acid	Nd		Nd			Nd		Nd		
XI	metolachlor	3.8	0.8	3.4	0.4	10	37	3	24	2	34
XII	hydroxymetolachlor	5.6	0.2	4.9	0.4	13	33	3	8.6	0.2	74
XIII	deschlorometolachlor	4.1	0.6	3.8	0.2	9	4.7	0.4	3.5	0.2	27
XIV	metolachlor morpholinone	Nd		Nd			45	3	Nd		
XV	metolachlor propanol	Nd		Nd			13	1	13	2	0
XVI	deschloroacetylmetolachlor	2.7	0.3	2.9	0.5	-9	Nd		1.7	1.1	
XVII	deschloroacetylmetolachlor propanol	Nd		Nd			Nd		Nd		
XVIII	metolachlor oxanilic acid	Nd		Nd			Nd		Nd		

(continued)

Table A.8 (Continued)

Site 8		Fall*					Spring†				
#	Compound	Raw	Std	Treated	Std	% Removal	Raw	Std	Treated	Std	% Removal
XIX	metolachlor ethane sulfonic acid	Nd		Nd			458	23	452	27	1
XX	acetochlor	0.77	0.43	0.70	0.03	9	15	2	11	4	29
XXI	hydroxyacetochlor	Nd		Nd			17	1	Nd		>97
XXII	deschloroacetochlor	Nd		Nd			Nd		Nd		
XXIII	acetochlor oxanilic acid	Nd		Nd			Nd		Nd		
XXIV	acetochlor ethane sulfonic acid	527	2	533	0	-1	Nd		Nd		
XXV	2-chloro-2'-ethyl-6'-methylacetanilide	5.7	0.3	19	1	-228	<0.7		1.2	0.3	<-42
XXVI	2-hydroxy-2'-ethyl-6'-methylacetanilide	Nd		Nd			6.7	1.4	17	2	-151
XXVII	2'-ethyl-6'-methylacetanilide	38	1	37	1	3	3.0	1.2	12	1	-315
XXVIII	2-ethyl-6-methylaniline	<25		<25			Nd		Nd		
XXIX	dimethenamid	2.4	0.4	2.3	0.4	4	Nd		Nd		
XXX	deschlorodimethenamid	5.5	0.5	2.6	0.5	52	Nd		Nd		
XXXI	atrazine	5.6	0.3	5.2	0.2	6	74	3	38	3	49
XXXII	desethyl atrazine	3.0	0.2	3.3	0.4	-11	5.9	1.1	8.0	0.8	-36
XXXIII	desisopropyl atrazine	Nd		Nd			6.3	0.1	8.2	0.4	-30
XXXIV	simazine	2.7	0.2	2.7	0.5	1	1.7	0.3	3.4	1.0	-95
XXXV	cyanazine	12	1	12	1	6	Nd		Nd		

NOTE: Water samples from groundwater

* Fall 2003 treatment train: ozone, chlorine, polyphosphate (ortho-poly blend)

† Spring 2004 treatment train: ozone, chlorine, polyphosphate (ortho-poly blend)

Table A.9

Site 9 measurements of target analytes in raw and treated drinking water and the percentage removal of each compound within the drinking water utility during Fall 2003 and Spring 2004. Concentrations are in ng/L.

Site 9		Fall[*]					Spring[†]				
#	Compound	Raw	Std	Treated	Std	% Removal	Raw	Std	Treated	Std	% Removal
I	alachlor	Nd		Nd			1.4	0.2	0.90	0.18	36
II	hydroxyalachlor	Nd		Nd			Nd		Nd		
III	deschloroalachlor	Nd		Nd			Nd		Nd		
IV	2-chloro-2'-6'-diethylacetanilide	4.8	0.7	5.0	0.3	-4	1.7	0.2	1.3	0.1	20
V	2-hydroxy-2'-6'-diethylacetanilide	Nd		Nd			65	1	74	2	-13
VI	2-hydroxy-2'-6'-diethyl-*N*-methylacetanilide	Nd		Nd			Nd		Nd		
VII	2'-6'-diethylacetanilide	26	1	25	1	5	6.2	0.3	6.3	0.8	-2
VIII	2,6-diethylaniline	Nd		Nd			Nd		Nd		
IX	alachlor oxanilic acid	64	0	59	1	8	82	13	46	4	44
X	alachlor ethane sulfonic acid	Nd		Nd			Nd		Nd		
XI	metolachlor	1.5	0.1	1.4	0.2	8	30	1	27	3	11
XII	hydroxymetolachlor	Nd		Nd			3.6	0.2	3.8	0.2	-8
XIII	deschlorometolachlor	4.0	0.1	3.6	0.1	10	4.5	0.3	4.3	0.4	4
XIV	metolachlor morpholinone	Nd		Nd			17	1	13	2	21
XV	metolachlor propanol	Nd		Nd			13	1	Nd		>96
XVI	deschloroacetylmetolachlor	1.6	0.3	2.3	0.2	-40	Nd		Nd		
XVII	deschloroacetylmetolachlor propanol	Nd		Nd			Nd		Nd		
XVIII	metolachlor oxanilic acid	49	1	50	0	-2	95	15	68	7	29

(continued)

Table A.9 (Continued)

Site 9		Fall*					Spring†				
#	Compound	Raw	Std	Treated	Std	% Removal	Raw	Std	Treated	Std	% Removal
XIX	metolachlor ethane sulfonic acid	Nd		Nd			Nd		Nd		
XX	acetochlor	Nd		Nd			100	1	94	2	6
XXI	hydroxyacetochlor	Nd		Nd			Nd		Nd		
XXII	deschloroacetochlor	Nd		Nd			2.2	0.4	1.9	0.1	13
XXIII	acetochlor oxanilic acid	31	1	27	1	12	80	13	53	12	34
XXIV	acetochlor ethane sulfonic acid	Nd		Nd			Nd		Nd		
XXV	2-chloro-2'-ethyl-6'-methylacetanilide	4.9	1.0	5.1	0.8	-2	3.1	1.3	Nd		>79
XXVI	2-hydroxy-2'-ethyl-6'-methylacetanilide	Nd		Nd			Nd		Nd		
XXVII	2'-ethyl-6'-methylacetanilide	34	1	30	1	12	6.4	0.4	7.9	0.6	-22
XXVIII	2-ethyl-6-methylaniline	Nd		Nd			Nd		Nd		
XXIX	dimethenamid	Nd		Nd			Nd		Nd		
XXX	deschlorodimethenamid	Nd		Nd			1.3	0.4	Nd		>77
XXXI	atrazine	3.1	0.1	3.0	0.1	1	56	1	47	7	16
XXXII	desethyl atrazine	Nd		Nd			9.4	0.6	6.7	0.5	29
XXXIII	desisopropyl atrazine	Nd		Nd			12	0.8	11	1	11
XXXIV	simazine	Nd		Nd			3.9	0.7	3.6	0.4	5
XXXV	cyanazine	Nd		Nd			Nd		Nd		

NOTE: Water samples from groundwater

* Fall 2003 treatment train: aeration, lime softening, chlorine, fluoride, filtration (dual media: anthracite/sand)

† Spring 2004 treatment train: lime softening, aeration, lime, chlorine, fluoride, filtration (dual media: anthracite/sand)

Table A.10

Site 10 measurements of target analytes in raw and treated drinking water and the percentage removal of each compound within the drinking water utility during Fall 2003 and Spring 2004. Concentrations are in ng/L.

Site 10		Fall*					Spring†				
#	Compound	Raw	Std	Treated	Std	% Removal	Raw	Std	Treated	Std	% Removal
I	alachlor	Nd		Nd			0.66	0.04	0.47	0.02	30
II	hydroxyalachlor	Nd		Nd			26	1	24	2	6
III	deschloroalachlor	Nd		Nd			Nd		Nd		
IV	2-chloro-2'-6'-diethylacetanilide	4.9	0.4	5.9	0.9	-21	0.58	0.21	0.51	0.11	12
V	2-hydroxy-2'-6'-diethylacetanilide	Nd		Nd			22	3	22	1	2
VI	2-hydroxy-2'-6'-diethyl-*N*-methylacetanilide	Nd		Nd			Nd		Nd		
VII	2'-6'-diethylacetanilide	22	1	24	1	-7	3.1	0.2	2.9	0.4	4
VIII	2,6-diethylaniline	Nd		Nd			Nd		Nd		
IX	alachlor oxanilic acid	22	1	21	1	5	98	1	93	3	5
X	alachlor ethane sulfonic acid						648	20	618	16	5
XI	metolachlor	5.7	0.4	5.9	0.5	-3	206	5	187	9	9
XII	hydroxymetolachlor	13	1	11	1	18	22	2	16	1	27
XIII	deschlorometolachlor	1.1	0.1	0.7	0.1	32	6.7	2.7	5.3	0.9	22
XIV	metolachlor morpholinone	Nd		Nd			29	4	22	0	26
XV	metolachlor propanol	Nd		Nd			16	3	14	1	9
XVI	deschloroacetylmetolachlor	0.18	0.03	0.34	0.14	-84	Nd		Nd		
XVII	deschloroacetylmetolachlor propanol	Nd		Nd			Nd		Nd		
XVIII	metolachlor oxanilic acid	33	1	34	1	-2	158	2	152	1	4

(continued)

Table A.10 (Continued)

Site 10		Fall*					Spring†				
#	Compound	Raw	Std	Treated	Std	% Removal	Raw	Std	Treated	Std	% Removal
XIX	metolachlor ethane sulfonic acid	1060	2	1010	1	5	1580	2	1520	2	3
XX	acetochlor	Nd		Nd			271	9	281	4	-3
XXI	hydroxyacetochlor	Nd		Nd			63	4	58	2	8
XXII	deschloroacetochlor	3.2	0.2	3.5	0.4	-8	11	2	9.8	1.1	7
XXIII	acetochlor oxanilic acid	37	1	35	1	5	466	19	458	25	2
XXIV	acetochlor ethane sulfonic acid						870	21	845	14	3
XXV	2-chloro-2'-ethyl-6'-methylacetanilide	9.3	0.6	9.4	0.9	-2	1.4	0.3	1.0	0.2	29
XXVI	2-hydroxy-2'-ethyl-6'-methylacetanilide	16	2	1	2	6	26	3	21	1	19
XXVII	2'-ethyl-6'-methylacetanilide	38	2	40	2	-5	6.6	0.8	6.6	0.7	1
XXVIII	2-ethyl-6-methylaniline	Nd		Nd			Nd		Nd		
XXIX	dimethenamid	Nd		Nd			66	3	67	2.3	-2
XXX	deschlorodimethenamid	Nd		Nd			Nd		Nd		
XXXI	atrazine	67	2	66	1	2	1350	9	1310	20	3
XXXII	desethyl atrazine	20	1	21	1	-7	162	5	148	8	9
XXXIII	desisopropyl atrazine	21	2	33	1	-56	38	4	36	1	5
XXXIV	simazine	4.6	0.4	7.2	0.5	-58	6.1	0.5	6.1	0.8	1
XXXV	cyanazine	9.1	0.8	9.6	0.8	-5	Nd		Nd		

NOTE: Water samples from groundwater

* Fall 2003 treatment train: softening, settling, chlorine, CO_2, sand filtration, phosphate, fluoride

† Spring 2004 treatment train: ammonia, aeration, lime, chlorine, CO_2, sand filtration, phosphate, fluoride

Table A.11
Site 11 measurements of target analytes in raw and treated drinking water and the percentage removal of each compound within the drinking water utility during Fall 2003 and Spring 2004. Concentrations are in ng/L.

Site 11		Fall*					Spring†				
#	Compound	Raw	Std	Treated	Std	% Removal	Raw	Std	Treated	Std	% Removal
I	alachlor	Nd		Nd			1.8	0.02	1.6	0.4	13
II	hydroxyalachlor	Nd		Nd			38	3	Nd		>78
III	deschloroalachlor	Nd		Nd			Nd		Nd		
IV	2-chloro-2'-6'-diethylacetanilide	6.0	0.3	6.7	0.7	-11	0.72	1.3	0.74	0.02	-2
V	2-hydroxy-2'-6'-diethylacetanilide	Nd		Nd			56	9	26	3	53
VI	2-hydroxy-2'-6'-diethyl-*N*-methylacetanilide	Nd		Nd			<11		<11		
VII	2'-6'-diethylacetanilide	32	1	38	1	-19	8.6	2.0	5.6	0.8	34
VIII	2,6-diethylaniline	Nd		Nd			Nd		Nd		
IX	alachlor oxanilic acid	Nd		Nd			50	0	42	2	15
X	alachlor ethane sulfonic acid	Nd		Nd			Nd		Nd		
XI	metolachlor	3.4	0.1	3.4	0.2	0	188	1	176	2	6
XII	hydroxymetolachlor	Nd		Nd			41	45	45	2	-10
XIII	deschlorometolachlor	3.2	0.1	3.3	0.1	-5	5.6	0.6	5.6	1.0	1
XIV	metolachlor morpholinone	Nd		Nd			24	6	29	2	-23
XV	metolachlor propanol	Nd		Nd			37	3	26	2	27
XVI	deschloroacetylmetolachlor	Nd		1.3	0.3	<-78	9.3	0.3	7.8	0.9	16
XVII	deschloroacetylmetolachlor propanol	Nd		Nd			13	1.2	4.7	0.7	65
XVIII	metolachlor oxanilic acid	<21		<21			28	3	25	1	12

(continued)

Table A.11 (Continued)

Site 11		Fall*					Spring†				
#	Compound	Raw	Std	Treated	Std	% Removal	Raw	Std	Treated	Std	% Removal
XIX	metolachlor ethane sulfonic acid	Nd		Nd			550	11	500	7	9
XX	acetochlor	Nd		Nd			87	5	81	6	7
XXI	hydroxyacetochlor	Nd		Nd			40	7	38	1	4
XXII	deschloroacetochlor	Nd		Nd			5.2	0.7	4.4	0.8	16
XXIII	acetochlor oxanilic acid	<20		<20			23	2	21	0	9
XXIV	acetochlor ethane sulfonic acid	Nd		Nd			Nd		Nd		
XXV	2-chloro-2'-ethyl-6'-methylacetanilide	58	1	5.3	0.8	91	11	5	7.0	0.7	35
XXVI	2-hydroxy-2'-ethyl-6'-methylacetanilide	Nd		Nd			98	22	31	1	68
XXVII	2'-ethyl-6'-methylacetanilide	39	1	43	1	-10	15	4	8.6	0.2	41
XXVIII	2-ethyl-6-methylaniline	<25		<25			Nd		Nd		
XXIX	dimethenamid	Nd		Nd			2.4	1.4	1.3	0.5	45
XXX	deschlorodimethenamid	Nd		Nd			Nd		Nd		
XXXI	atrazine	16	1	15	1	7	1070	36	988	39	8
XXXII	desethyl atrazine	32	45	6.7	0.7	79	360	41	150	6	58
XXXIII	desisopropyl atrazine	Nd		Nd			199	15	46	2	77
XXXIV	simazine	5.4	0.2	5.4	0.5	1	201	25	193	13	4
XXXV	cyanazine	Nd		Nd			Nd		Nd		

NOTE: Water samples from surface water

* Fall 2003 treatment train: ferric chloride, chlorine, ammonia, lime, aluminum chlorohydrate, anthracite/sand filtration

† Spring 2004 treatment train: ferric chloride, pDADMAC, chlorine, ammonia, lime, aluminum chlorohydrate, anthracite/sand filtration

Table A.12

Site 12 measurements of target analytes in raw and treated drinking water and the percentage removal of each compound within the drinking water utility during Fall 2003 and Spring 2004. Concentrations are in ng/L.

Site 12		Fall*					Spring†				
#	Compound	Raw	Std	Treated	Std	% Removal	Raw	Std	Treated	Std	% Removal
I	alachlor	Nd		Nd			Nd		Nd		
II	hydroxyalachlor	Nd		Nd			28	12	29	0.1	-6
III	deschloroalachlor	Nd		Nd			Nd		Nd		
IV	2-chloro-2'-6'-diethylacetanilide	8.6	0.3	5.4	0.3	37	0.43	0.02	0.99	0.09	-128
V	2-hydroxy-2'-6'-diethylacetanilide	Nd		Nd			54	5	50	2	8
VI	2-hydroxy-2'-6'-diethyl-*N*-methylacetanilide	Nd		Nd			<11		Nd		
VII	2'-6'-diethylacetanilide	30	2	24	2	20	12	0.4	15	1	-23
VIII	2,6-diethylaniline	Nd		Nd			Nd		Nd		
IX	alachlor oxanilic acid	Nd		Nd			30	1	27	1	8
X	alachlor ethane sulfonic acid	Nd		Nd			Nd		Nd		
XI	metolachlor	5.3	0.1	5.0	0.3	5	169	3	161	11	5
XII	hydroxymetolachlor	Nd		Nd			41	1	45	6	-11
XIII	deschlorometolachlor	3.5	0.1	3.2	0.0	9	32	3	30	1	7
XIV	metolachlor morpholinone	Nd		Nd			22	1	18	1	16
XV	metolachlor propanol	Nd		Nd			30	3	26	3	15
XVI	deschloroacetylmetolachlor	Nd		0.72	0.12	<-60	39	5	35	3	11
XVII	deschloroacetylmetolachlor propanol	Nd		Nd			17	1	22	2	-29
XVIII	metolachlor oxanilic acid	22	1	25	1	-17	22	1	20	3	8

(continued)

Table A.12 (Continued)

Site 12		Fall*					Spring†				
#	Compound	Raw	Std	Treated	Std	% Removal	Raw	Std	Treated	Std	% Removal
XIX	metolachlor ethane sulfonic acid	Nd		Nd			607	24	578	0	5
XX	acetochlor	Nd		Nd			27	1	20	2	25
XXI	hydroxyacetochlor	Nd		Nd			38	3	57	2	-51
XXII	deschloroacetochlor	Nd		Nd			35	2	31	3	12
XXIII	acetochlor oxanilic acid	21	1	23	1	-8	23	0	20	1	11
XXIV	acetochlor ethane sulfonic acid	Nd		Nd			Nd		Nd		
XXV	2-chloro-2'-ethyl-6'-methylacetanilide	71	2	3.9	0.4	95	12	1	6.6	0.8	45
XXVI	2-hydroxy-2'-ethyl-6'-methylacetanilide	Nd		Nd			83	5	67	1	20
XXVII	2'-ethyl-6'-methylacetanilide	39	1	33	1	15	25	2	17	1	30
XXVIII	2-ethyl-6-methylaniline	<25		<25			Nd		Nd		
XXIX	dimethenamid	Nd		Nd			0.55	0.25	0.35	0.08	36
XXX	deschlorodimethenamid	Nd		Nd			Nd		Nd		
XXXI	atrazine	17	1	14	1	15	837	18	819	38	2
XXXII	desethyl atrazine	6.1	1.0	5.2	0.3	15	100	5	83	2	18
XXXIII	desisopropyl atrazine	Nd		Nd			28	1	21	2	25
XXXIV	simazine	4.9	0.5	3.8	0.1	22	148	4.8	145	4	2
XXXV	cyanazine	Nd		Nd			Nd		Nd		

NOTE: Water samples from surface water in spring, in fall 3:1 ratio of surface water to riverbank filtered water

* Fall 2003 treatment train: ferric chloride, lime, chlorine, dual media filtration, ammonia

† Spring 2004 treatment train: ferric chloride, pDADMAC, chlorine, ammonia, lime, dual media filtration

Table A.13

Comparison of neutral analytes in raw and treated drinking water at Site 3 for extractions conducted immediately upon sample arrival and extractions conducted after samples were stored for 28 days at 6 °C. Concentrations are in ng/L.

Site 3		Raw			Finished		
#	Compound	0 Days	28 Days	% Change	0 Days	28 Days	% Change
I	alachlor	9.2	8.3	-10	7.9	6.6	-16
II	hydroxyalachlor	19	18	-5	Nd	Nd	
III	deschloroalachlor	Nd	Nd		Nd	Nd	
IV	2-chloro-2'-6'-diethylacetanilide	16	13	-19	11	11	0
V	2-hydroxy-2'-6'-diethylacetanilide	5.1	5.7	12	9.2	8.1	-12
VI	2-hydroxy-2'-6'-diethyl-*N*-methylacetanilide	Nd	Nd		Nd	Nd	
VII	2'-6'-diethylacetanilide	43	43	0	38	37	-3
VIII	2,6-diethylaniline	Nd	Nd		Nd	Nd	
XI	metolachlor	20	21	5	16	16	0
XII	hydroxymetolachlor	33	32	-3	25	24	-4
XIII	deschlorometolachlor	3.6	4.6	28	2.8	3.2	14
XIV	metolachlor morpholinone	Nd	Nd		Nd	Nd	
XV	metolachlor propanol	Nd	Nd		Nd	Nd	
XVI	deschloroacetylmetolachlor	0.89	0.8	-10	2.8	2.1	-25
XVII	deschloroacetylmetolachlor propanol	Nd	Nd		Nd	Nd	
XX	acetochlor	6.5	5.9	-9	4.5	4.5	0
XXI	hydroxyacetochlor	39	39	0	32	30	-6
XXII	deschloroacetochlor	27	27	0	20	21	5
XXV	2-chloro-2'-ethyl-6'-methylacetanilide	43	45	5	37	35	-5
XXVI	2-hydroxy-2'-ethyl-6'-methylacetanilide	2.7	3.2	19	9	8.9	-1
XXVII	2'-ethyl-6'-methylacetanilide	57	57	0	57	56	-2
XXVIII	2-ethyl-6-methylaniline	<25	<25		<25	<25	
XXIX	dimethenamid	6.8	6.5	-4	Nd	Nd	
XXX	deschlorodimethenamid	14	12	-14	25	22	-12
XXXI	atrazine	133	131	-2	106	107	1
XXXII	desethyl atrazine	44	45	2	44	45	2
XXXIII	desisopropyl atrazine	10	11	10	11	10	-9
XXXIV	simazine	12	13	8	12	13	8
XXXV	cyanazine	Nd	Nd		Nd	Nd	

Table A.14

Comparison of neutral analytes in raw and treated drinking water at Site 4 for extractions conducted immediately upon sample arrival and extractions conducted after samples were stored for 28 days at 6 °C. Concentrations are in ng/L.

Site 4		Raw			Finished		
#	Compound	0 Days	28 Days	% Change	0 Days	28 Days	% Change
I	alachlor	6.8	6.6	-3	6.6	6.5	-2
II	hydroxyalachlor	Nd	Nd		Nd	Nd	
III	deschloroalachlor	Nd	Nd		Nd	Nd	
IV	2-chloro-2'-6'-diethylacetanilide	5.5	5.1	-7	4.7	4.9	4
V	2-hydroxy-2'-6'-diethylacetanilide	Nd	Nd		Nd	Nd	
VI	2-hydroxy-2'-6'-diethyl-*N*-methylacetanilide	Nd	Nd		Nd	Nd	
VII	2'-6'-diethylacetanilide	24	25	4	26	26	0
VIII	2,6-diethylaniline	Nd	Nd		Nd	Nd	
XI	metolachlor	16	16	0	16	16	0
XII	hydroxymetolachlor	26	26	0	24	26	8
XIII	deschlorometolachlor	3	3.2	7	2.9	2.9	0
XIV	metolachlor morpholinone	Nd	Nd		Nd	Nd	
XV	metolachlor propanol	Nd	Nd		Nd	Nd	
XVI	deschloroacetylmetolachlor	1.2	1.4	17	2.5	2.5	0
XVII	deschloroacetylmetolachlor propanol	Nd	Nd		Nd	Nd	
XX	acetochlor	6.9	6.4	-7	6.6	6.4	-3
XXI	hydroxyacetochlor	40	37	-8	38	38	0
XXII	deschloroacetochlor	13	13	0	14	14	0
XXV	2-chloro-2'-ethyl-6'-methylacetanilide	16	16	0	22	21	-5
XXVI	2-hydroxy-2'-ethyl-6'-methylacetanilide	Nd	Nd		Nd	Nd	
XXVII	2'-ethyl-6'-methylacetanilide	36	36	0	35	35	0
XXVIII	2-ethyl-6-methylaniline	<25	<25		<25	<25	
XXIX	dimethenamid	3.4	3.7	9	2.7	2.5	-7
XXX	deschlorodimethenamid	4.9	3.8	-22	2	2.1	5
XXXI	atrazine	135	135	0	131	132	1
XXXII	desethyl atrazine	33	33	0	31	30	-3
XXXIII	desisopropyl atrazine	11	11	0	10	10	0
XXXIV	simazine	10	11	10	9.8	9.5	-3
XXXV	cyanazine	Nd	Nd		Nd	Nd	

Table A.15

Comparison of neutral analytes in raw and treated drinking water at Site 5 for extractions conducted immediately upon sample arrival and extractions conducted after samples were stored for 28 days at 6 °C. Concentrations are in ng/L.

Site 5		Raw			Finished		
#	Compound	0 Days	28 Days	% Change	0 Days	28 Days	% Change
I	alachlor	5.7	4.7	-18	5	4.6	-8
II	hydroxyalachlor	Nd	Nd		Nd	Nd	
III	deschloroalachlor	Nd	Nd		Nd	Nd	
IV	2-chloro-2'-6'-diethylacetanilide	11	9.8	-11	8.5	8	-6
V	2-hydroxy-2'-6'-diethylacetanilide	Nd	Nd		Nd	Nd	
VI	2-hydroxy-2'-6'-diethyl-*N*-methylacetanilide	Nd	Nd		Nd	Nd	
VII	2'-6'-diethylacetanilide	35	37	6	25	23	-8
VIII	2,6-diethylaniline	Nd	Nd		Nd	Nd	
XI	metolachlor	30	28	-7	24	25	4
XII	hydroxymetolachlor	29	29	0	28	27	-4
XIII	deschlorometolachlor	1.0	1.4	40	2.5	3	20
XIV	metolachlor morpholinone	Nd	Nd		Nd	Nd	
XV	metolachlor propanol	Nd	Nd		Nd	Nd	
XVI	deschloroacetylmetolachlor	2.4	2.8	17	1.6	1.1	-31
XVII	deschloroacetylmetolachlor propanol	Nd	Nd		Nd	Nd	
XX	acetochlor	4.7	4.1	-13	4.5	4.1	-9
XXI	hydroxyacetochlor	22	24	9	19	21	11
XXII	deschloroacetochlor	Nd	Nd		8	7.2	-10
XXV	2-chloro-2'-ethyl-6'-methylacetanilide	28	26	-7	26	25	-4
XXVI	2-hydroxy-2'-ethyl-6'-methylacetanilide	Nd	Nd		Nd	Nd	
XXVII	2'-ethyl-6'-methylacetanilide	53	54	2	37	39	5
XXVIII	2-ethyl-6-methylaniline	<25	<25		<25	<25	
XXIX	dimethenamid	2.5	2.2	-12	2.2	2	-9
XXX	deschlorodimethenamid	4.7	4.3	-9	2.5	1.9	-24
XXXI	atrazine	201	198	-1	151	152	1
XXXII	desethyl atrazine	32	33	3	46	43	-7
XXXIII	desisopropyl atrazine	7.4	8.2	11	9.9	8.9	-10
XXXIV	simazine	8.6	7.7	-10	6.3	6	-5
XXXV	cyanazine	8.7	8.4	-3	7.9	8.2	4

Table A.16

Comparison of neutral analytes in raw and treated drinking water at Site 7 for extractions conducted immediately upon sample arrival and extractions conducted after samples were stored for 28 days at 6 °C. Concentrations are in ng/L.

Site 7		Raw			Finished		
#	Compound	0 Days	28 Days	% Change	0 Days	28 Days	% Change
I	alachlor	1.3	1.4	8	1.2	1.6	33
II	hydroxyalachlor	11	11	0	5.2	4.1	-21
III	deschloroalachlor	Nd	Nd		Nd	Nd	
IV	2-chloro-2'-6'-diethylacetanilide	6.3	6.2	-2	1.7	1.6	-6
V	2-hydroxy-2'-6'-diethylacetanilide	Nd	Nd		Nd	Nd	
VI	2-hydroxy-2'-6'-diethyl-*N*-methylacetanilide	Nd	Nd		Nd	Nd	
VII	2'-6'-diethylacetanilide	24	24	0	22	22	0
VIII	2,6-diethylaniline	Nd	Nd		Nd	Nd	
XI	metolachlor	26	26	0	24	23	-4
XII	hydroxymetolachlor	14	15	7	11	11	0
XIII	deschlorometolachlor	6	6.1	2	5.2	5.5	6
XIV	metolachlor morpholinone	12	11	-8	8.8	8.8	0
XV	metolachlor propanol	Nd	Nd		Nd	Nd	
XVI	deschloroacetylmetolachlor	5.1	5.5	8	5.6	5.5	-2
XVII	deschloroacetylmetolachlor propanol	Nd	Nd		Nd	Nd	
XX	acetochlor	0.93	0.9	-3	0.51	0.6	18
XXI	hydroxyacetochlor	17	17	0	13	14	8
XXII	deschloroacetochlor	Nd	Nd		Nd	Nd	
XXV	2-chloro-2'-ethyl-6'-methylacetanilide	11	11	0	8.8	8.1	-8
XXVI	2-hydroxy-2'-ethyl-6'-methylacetanilide	Nd	Nd		Nd	Nd	
XXVII	2'-ethyl-6'-methylacetanilide	31	32	3	28	28	0
XXVIII	2-ethyl-6-methylaniline	<25	<25		<25	<25	
XXIX	dimethenamid	1.4	1.8	29	1.5	1.6	7
XXX	deschlorodimethenamid	2.1	2.5	19	2.2	2	-9
XXXI	atrazine	186	185	-1	131	132	1
XXXII	desethyl atrazine	56	56	0	26	28	8
XXXIII	desisopropyl atrazine	14	14	0	9	9.6	7
XXXIV	simazine	49	49	0	32	32	0
XXXV	cyanazine	11	11	0	9.3	11	18

Table A.17

Comparison of neutral analytes in raw and treated drinking water at Site 8 for extractions conducted immediately upon sample arrival and extractions conducted after samples were stored for 28 days at 6 °C. Concentrations are in ng/L.

Site 8		Raw			Finished		
#	Compound	0 Days	28 Days	% Change	0 Days	28 Days	% Change
I	alachlor	1.7	1.6	-6	1.5	1.9	27
II	hydroxyalachlor	Nd	Nd		Nd	Nd	
III	deschloroalachlor	Nd	Nd		Nd	Nd	
IV	2-chloro-2'-6'-diethylacetanilide	4.9	5	2	6.1	6.7	10
V	2-hydroxy-2'-6'-diethylacetanilide	Nd	Nd		Nd	Nd	
VI	2-hydroxy-2'-6'-diethyl-*N*-methylacetanilide	Nd	Nd		Nd	Nd	
VII	2'-6'-diethylacetanilide	28	27	-4	28	28	0
VIII	2,6-diethylaniline	Nd	Nd		Nd	Nd	
XI	metolachlor	3.8	3.5	-8	3.4	3.3	-3
XII	hydroxymetolachlor	5.6	5.6	0	4.9	4.3	-12
XIII	deschlorometolachlor	4.1	4	-2	3.8	3.5	-8
XIV	metolachlor morpholinone	Nd	Nd		Nd	Nd	
XV	metolachlor propanol	Nd	Nd		Nd	Nd	
XVI	deschloroacetylmetolachlor	2.7	2.5	-7	2.9	2	-31
XVII	deschloroacetylmetolachlor propanol	Nd	Nd		Nd	Nd	
XX	acetochlor	0.77	0.7	-9	0.7	0.7	0
XXI	hydroxyacetochlor	Nd	Nd		Nd	Nd	
XXII	deschloroacetochlor	Nd	Nd		Nd	Nd	
XXV	2-chloro-2'-ethyl-6'-methylacetanilide	5.7	5.6	-2	19	19	0
XXVI	2-hydroxy-2'-ethyl-6'-methylacetanilide	Nd	Nd		Nd	Nd	
XXVII	2'-ethyl-6'-methylacetanilide	38	37	-3	37	38	3
XXVIII	2-ethyl-6-methylaniline	<25	<25		<25	<25	
XXIX	dimethenamid	2.4	2.4	0	2.3	2.2	-4
XXX	deschlorodimethenamid	5.5	5.6	2	2.6	2	-23
XXXI	atrazine	5.6	5.3	-5	5.2	5.1	-2
XXXII	desethyl atrazine	3	2.7	-10	3.3	3.2	-3
XXXIII	desisopropyl atrazine	Nd	Nd		Nd	Nd	
XXXIV	simazine	2.7	2.4	-11	2.7	2.4	-11
XXXV	cyanazine	12	11	-8	12	11	-8

Table A.18

Comparison of neutral analytes in raw and treated drinking water at Site 9 for extractions conducted immediately upon sample arrival and extractions conducted after samples were stored for 28 days at 6 °C. Concentrations are in ng/L.

Site 9		Raw			Finished		
#	Compound	0 Days	28 Days	% Change	0 Days	28 Days	% Change
I	alachlor	Nd	Nd		Nd	Nd	
II	hydroxyalachlor	Nd	Nd		Nd	Nd	
III	deschloroalachlor	Nd	Nd		Nd	Nd	
IV	2-chloro-2'-6'-diethylacetanilide	4.8	5	4	5	4.5	-10
V	2-hydroxy-2'-6'-diethylacetanilide	Nd	Nd		Nd	Nd	
VI	2-hydroxy-2'-6'-diethyl-*N*-methylacetanilide	Nd	Nd		Nd	Nd	
VII	2'-6'-diethylacetanilide	26	26	0	25	25	0
VIII	2,6-diethylaniline	Nd	Nd		Nd	Nd	
XI	metolachlor	1.5	1.4	-7	1.4	1.4	0
XII	hydroxymetolachlor	Nd	Nd		Nd	Nd	
XIII	deschlorometolachlor	4	4.7	18	3.6	3.8	6
XIV	metolachlor morpholinone	Nd	Nd		Nd	Nd	
XV	metolachlor propanol	Nd	Nd		Nd	Nd	
XVI	deschloroacetylmetolachlor	1.6	1.2	-25	2.3	2.1	-9
XVII	deschloroacetylmetolachlor propanol	Nd	Nd		Nd	Nd	
XX	acetochlor	Nd	Nd		Nd	Nd	
XXI	hydroxyacetochlor	Nd	Nd		Nd	Nd	
XXII	deschloroacetochlor	Nd	Nd		Nd	Nd	
XXV	2-chloro-2'-ethyl-6'-methylacetanilide	4.9	5.2	6	5.1	5.1	0
XXVI	2-hydroxy-2'-ethyl-6'-methylacetanilide	Nd	Nd		Nd	Nd	
XXVII	2'-ethyl-6'-methylacetanilide	34	33	-3	30	30	0
XXVIII	2-ethyl-6-methylaniline	Nd	Nd		Nd	Nd	
XXIX	dimethenamid	Nd	Nd		Nd	Nd	
XXX	deschlorodimethenamid	Nd	Nd		Nd	Nd	
XXXI	atrazine	3.1	3.1	0	3	3	0
XXXII	desethyl atrazine	Nd	Nd		Nd	Nd	
XXXIII	desisopropyl atrazine	Nd	Nd		Nd	Nd	
XXXIV	simazine	Nd	Nd		Nd	Nd	
XXXV	cyanazine	Nd	Nd		Nd	Nd	

APPENDIX B
HANDOUTS FOR UTILITY PERSONNEL

INSTRUCTIONS FOR SAMPLING

Cooler contains four 2-L glass bottles

- 2 bottles contain preservatives (orange tape around top of bottle)
  - 1 bottle is marked for raw water, 1 bottle is marked for finished water
- 2 bottles are empty
  - 1 bottle is marked for raw water, 1 bottle is marked for finished water

When filling the bottles please <u>do not</u> wash out the preservatives

- Overfilling of the bottles is not necessary
- A small amount of headspace at the top of the bottle is fine

2 bottles, already marked, should be used for the raw (source) water

- 1 bottle with preservatives (orange tape)
- 1 empty bottle

2 bottles, already marked, should be used for the finished (treated) water

- 1 bottle with preservatives (orange tape)
- 1 empty bottle

Once the samples are obtained, please cap them tightly, and wrap the caps with the enclosed lab tape

For the samples with preservatives, please agitate the bottles by hand until all preservatives are dissolved

After taking samples please mark metal tags on handles (with a ballpoint pen) to indicate the date and time of the sample

Fill out questionnaire

Repack bottles with bubble wrap, replace the instructions and the questionnaire in the cooler

Please secure the lid of the cooler with some of the enclosed strapping tape by wrapping strapping tape all the way around the midway point of the cooler

Please remove original shipping label and replace with enclosed shipping label for FedEx overnight delivery

Thank you very much for your help!

Any questions or problems please contact:
Michelle Hladik
Email: Hladik@jhu.edu
Lab: 410-516-7807

QUESTIONNAIRE

Utility location

____

Source of water (groundwater, surface water, mix). If a mix, is the ratio known; if surface water where is water obtained from (a side-channel or the middle of the river)

____

____

____

____

Describe the treatment train at your facility at *the time of sampling*. Please be specific as to which coagulant and oxidant are used. Also note if GAC or PAC was in use at the time the samples were obtained.

____

____

____

____

What is flow of water in the treatment plant (units: millions of gallons per day)

____

Once results are obtained we will provide your facility with the results. We will keep any results from your facility anonymous if requested.

Michelle L. Hladik, Ph.D. Candidate
Johns Hopkins University
Department of Geography and Environmental Engineering
313 Ames Hall
3400 N. Charles St.
Baltimore, MD 21218

Phone: 410-516-6579
Lab: 410-516-7807
E-mail: hladik@jhu.edu

REFERENCES

Acero, J.L., K. Stemmler, and U. von Gunten. 2000. Degradation Kinetics of Atrazine and Its Degradation Products With Ozone and OH Radicals: A Predictive Tool for Drinking Water Treatment. *Environ. Sci. Technol.*, 34:591–597.

Adams, C.D., and S.J. Randtke. 1992. Ozonation Byproducts of Atrazine in Synthetic and Natural Waters. *Environ. Sci. Technol.*, 26(11):2218–2227.

Adams, C.D., and T.L. Watson. 1996. Treatability of s-Triazine Herbicide Metabolites Using Powdered Activated Carbon. *Jour. Environ. Eng.*, 327–330.

Aga, D.S., E.M. Thurman, M.E. Yockel, L.R. Zimmerman, and T.D. Williams. 1996. Identification of a New Sulfidic Acid Metabolite of Metolachlor in Soil. *Environ. Sci. Technol.*, 30:592–597.

APHA, AWWA, and WEF (American Public Health Association, American Water Works Association, and Water Environment Federation). 1992. *Standard Methods for the Examination of Water and Wastewater.* 18th ed. Washington, D.C.: APHA.

ARP (Acetochlor Registration Partnership) 2002. Internet Web site [Online]. Available: <http://www.arpinfo.com/>. [cited September 11, 2002]

Aspelin, A.L., and A.H. Grube. 1999. United States Environmental Protection Agency, Office of Pesticide Programs. USEPA Pesticides Industry Sales and Usage: 1996 and 1997 Market Estimates [Online]. Available: <http://www.epa.gov/oppbead1/pestsales/ 97pestsales/table8.htm>. [cited May 14, 2002]

Baker, D.B., R.J. Rodney, S.A. Adams, and C. Macomber. 1993. Immunoassay Screens for Alachlor in Rural Wells: False Positives and an Alachlor Soil Metabolite. *Environ. Sci. Technol.*, 27(3):562–564.

Barrett, M.R. 1996. The Environmental Impact of Pesticide Degradates in Groundwater. In *Herbicide Metabolites in Surface Water and Groundwater.* Edited by M.T. Meyer and E.M. Thurman. Washington, D.C.: American Chemical Society.

Battaglin, W.A., E.T. Furlong, M.R. Burkhardt, and C.J. Peter. 2000. Occurrence of Sulfonylurea, Sulfonamide, Imidazolinone, and Other Herbicides in Rivers, Reservoirs and Ground Water in the Midwestern United States, 1998. *Sci. Total Environ.*, 248:123–133.

Beltrán, F., B. Acedo, and J. Rivas. 1999a. Use of Ozone and Hydrogen Peroxide to Remove Alachlor From Surface Water. *Bull. Environ. Contam. Toxicol.*, 63:9–14.

Beltrán, F., B. Acedo, and J. Rivas. 1999b. Use of Ozone to Remove Alachlor From Surface Water. *Bull. Environ. Contam. Toxicol.*, 62:324–329.

Beltrán, F., J. Rivas, and B. Acedo. 1999. Atrazine Removal by Ozonation Processes in Surface Waters. *Jour. Environ. Sci. Health*, B34(3):449–468.

Beltrán, F.J., B. Acedo, J. Rivas, and P. Alvárez. 2000. Comparison of Different Treatments of Alachlor Removal From Water. *Bull. Environ. Contam. Toxicol.*, 65:668–674.

Boyd, R.A. 2000. Herbicide and Herbicide Degradates in Shallow Groundwater and the Cedar River Near a Municipal Well Field, Cedar Rapids, Iowa. *Sci. Total Environ.*, 248:241–253.

Carlson, D.L. 2003. Environmental Transformations of Chloroacetamide Herbicides: Hydrolysis and Reactions With Iron Pyrite. Ph.D. diss., Johns Hopkins University, Baltimore, Md.

Carlson, D.L., M.M. McGuire, A.L. Roberts, and D.H. Fairbrother. 2003. Influence of Surface Composition on the Kinetics of Alachlor Reduction by Iron Pyrite. *Environ. Sci. Technol.*, 37(11):2394–2399.

Chan, W.F., and R.A. Larson. 1991. Formation of Mutagens From the Aqueous Reactions of Ozone With Anilines. *Water Res.*, 25(12):1529–1538.

Chesters, G., G.V. Simsiman, J. Levy, B.J. Alhajjar, R.N. Fathulla, and J.M. Harkin. 1989. Environmental Fate of Alachlor and Metolachlor. *Rev. Environ. Contam. Toxicol.*, 110:1–74.

Chiron, S., J. Abian, M. Ferrer, F. Sanchez-Baeza, A. Messeguer, and D. Barceló. 1995. Comparative Photodegradation Rates of Alachlor and Bentazone in Natural Water and Determination of Breakdown Products. *Environ. Toxicol. Chem.*, 14(8):1287–1298.

Christen, K. 2002. Prioritizing Drinking Water Contaminants. *Environ. Sci. Technol.*, 36(17):343A.

Dearfield, K.L., N.E. McCarroll, A. Protzel, H.F. Stack, M.A. Jackson, and M.D. Waters. 1999. A Survey of EPA/OPP and Open Literature on Selected Pesticide Chemicals II. Mutagenicity and Carcinogenicity of Selected Chloroacetanilides and Related Compounds. *Mutat. Res.*, 443:183–221.

EPA (U.S. Environmental Protection Agency). 2002 Acetochlor [Online]. Office of Prevention, Pesticides and Toxic Substances. Available: <http://www.epa.gov/oppefed1/aceto/index.htm>. [cited May 24, 2002]

EPA (U.S. Environmental Protection Agency). 1995. Metolachlor: Registration Eligibility Decision (RED) Fact Sheet [Online]. Office of Prevention, Pesticides and Toxic Substances. Available: <http://www.epa.gov/oppsrrd1/REDs/factsheets/0001fact.pdf>. [cited May 24, 2002]

EPA (U.S. Environmental Protection Agency). 1998a. Alachlor: Registration Eligibility Decision (RED) Fact Sheet [Online]. Office of Prevention, Pesticides and Toxic Substances. Available: <http://www.epa.gov/REDs/factsheets/0063fact.pdf>. [cited May 24, 2002]

EPA (U.S. Environmental Protection Agency). 1998b. Announcement of the Drinking Water Contaminant Candidate List [Online]. Available: <http://www.epa.gov/OGWDW/ccl/ccl_fr.pdf>. [cited July 31, 2002]

EPA (U.S. Environmental Protection Agency). 1999. Chemicals Evaluated for Carcinogenic Potential [Online]. Office of Pesticide Programs, Health Effects Division. Available: <http://www.epa.gov/pesticides/carlist/>. [cited July 31, 2002]

EPA (U.S. Environmental Protection Agency). 2000. Status of Chemicals in Special Review [Online]. Available: <http://www.epa.gov/oppsrrd1/docs/sr00status.pdf>. [cited September 23, 2004]

EPA (U.S. Environmental Protection Agency). 2001a. The Incorporation of Water Treatment Effects on Pesticide Removal and Transformations in Food Quality Protection Act (FQPA) Drinking Water Assessments [Online]. Office of Pesticide Programs. Available: <http://www.epa.gov/pesticides/trac/science/water_treatment.pdf>. [cited April 2002]

EPA (U.S. Environmental Protection Agency). 2001b. National Primary Drinking Water Standards [Online]. Available: <http://www.epa.gov/safewater/mcl.html>. [cited July 31, 2002]

EPA (U.S. Environmental Protection Agency). 2002a. Atrazine Tolerance Reassessment and Reregistration [Online]. Office of Prevention, Pesticides and Toxic Substances. Available: <http://www.epa.gov/pesticides/reregistration/atrazine/>. [cited May 24, 2002]

EPA (U.S. Environmental Protection Agency). 2002b. The Grouping of a Series of Triazine Pesticides Based on a Common Mechanism of Toxicity [Online]. Office of Pesticide Programs, Health Effects Division. Available: <http://www.epa.gov/oppsrrd1/cumulative/triazines/triazinescommonmech.pdf>. [cited May 24, 2002]

Eykholt, G.R., and D.T. Davenport. 1998. Dechlorination of the Chloroacetanilide Herbicides Alachlor and Metolachlor by Iron Metal. *Environ. Sci. Technol.*, 32:1482–1487.

Feng, P.C.C. 1991. Soil Transformation of Acetochlor via Glutathione Conjugation. *Pestic. Biochem. Physiol.*, 40:136–142.

Ferrer, I., E.M. Thurman, and D. Barceló. 1997. Identification of Ionic Chloroacetanilide-Herbicide Metabolites in Surface Water and Groundwater by HPLC/MS Using Negative Ion Spray. *Anal. Chem.*, 69:4547–4553.

Ferrer, I., E.M. Thurman, and D. Barceló. 2000. First LC/MS Determination of Cyanazine Amide, Cyanazine Acid, and Cyanazine in Ground Water Samples. *Environ. Sci. Technol.*, 34(4):714–718.

Field, J.A., and E.M. Thurman. 1996. Glutathione Conjugation and Contaminant Transformation. *Environ. Sci. Technol.*, 30(5):1413–1418.

Galassi, S., A. Provini, S. Mangiapan, and E. Benfenati. 1996. Alachlor and Its Metabolites in Surface Water. *Chemosphere*, 32(2):229–237.

Graham, W.H., D.W. Graham, F. Denoyelles Jr., V.H. Smith, C.K. Larive, and E.M. Thurman. 1999. Metolachlor and Alachlor Breakdown Product Formation Patterns in Aquatic Field Mesocosms. *Environ. Sci. Technol.*, 33:4471–4476.

Gustafson, D.I., K.H. Carr, D.B. Carson, J.D. Fuhrman, A.G. Hackett, T.J. Hoogheem, V.L. Snoeyink, M. Curry, B. Heijman, S. Chen, P. Hertl, and I. van Wesenbeeck. 2003. Activated Carbon Adsorption of Chloroacetanilide Herbicides and Their Degradation Products From Surface Water Supplies. *Jour. Water SRT-Aqua*, 52(6):443–454.

Hapeman-Somich, C.J., Z. Gui-Ming, W.R. Lusby, M.T. Muldoon, and R. Waters. 1992. Aqueous Ozonation of Atrazine. Products Identification and Description of the Degradation Pathway. *Jour. Agric. Food Chem.*, 40:2294–2298.

Hayes, T.B., A. Collins, M. Lee, M. Mendoza, N. Noriega, A.A. Stuart, and A. Vonk. 2002. Hermaphroditic, Demasculinized Frogs After Exposure to the Herbicide Atrazine at Low Ecologically Relevant Doses. *Proc. Natl. Acad. Sci.*, 99:5476–5480.

Heydens, W.F., A.G.E. Wilson, L.D. Kier, H. Lau, D.C. Thake, and M.A. Martens. 1999. An Evaluation of the Carcinogenic Potential of the Herbicide Alachlor to Man. *Human Exp. Toxicol.*, 18:363–391.

Heydens, W.F., A.G.E. Wilson, L.J. Kraus, W.E. Hopkins II, and K.J. Hotz. 2000. Ethane Sulfonate Metabolite of Alachlor: Assessment of Oncogenic Potential Based on Metabolic and Mechanistic Considerations. *Toxicol. Sci.*, 55:36–43.

Hill, A.B., P.R. Jefferies, G.B. Quistad, and J.E. Casida. 1997. Dialkylquinoneimine Metabolites of Chloroacetanilide Herbicides Induce Sister Chromatid Exchanges in Cultured Human Lymphocytes. *Mutat. Res.*, 395:159–171.

Hladik, M.L. 2005. Occurrence of Neutral Chloroacetamide Herbicide Degradates in Natural Waters and Their Removal During Drinking Water Treatment. Ph.D. diss., Johns Hopkins University, Baltimore, Md.

Hladik, M.L., A.L. Roberts, and E.J. Bouwer. 2005. Removal of Neutral Chloroacetamide Herbicide Degradates During Simulated Unit Processes for Drinking Water Treatment. *Water Res.*, 39(20):5033–5044.

Hladik, M.L., J.J. Hsiao, and A.L. Roberts. 2005. Are Neutral Chloroacetamide Degradates "Emerging Contaminants" of Potential Concern? Analysis and Occurrence in the Upper Chesapeake Bay. *Environ. Sci. Technol.*, 39(17):6561–6574.

Hogenboom, A.C., W.M.A. Niessen, and U.A.T. Brinkman. 2000. Characterization of Photodegradation Products of Alachlor in Water by On-line Solid-Phase Extraction Liquid Chromatography Combined With Tandem Mass Spectrometry and Orthogonal-Acceleration Time-of-Flight Mass Spectrometry. *Rapid. Commun. Mass Spectrom.*, 14:1914–1924.

Hostetler, K.A., and E.M. Thurman. 2000. Determination of Chloroacetanilide Herbicide Metabolites in Water Using High-Performance Liquid Chromatography-Diode Array Detection and High-Performance Liquid Chromatography/Mass Spectrometry. *Sci. Total Environ.*, 248:147–155.

Hudson, H.E. 1981. *Water Clarification Processes Practical Design and Evaluation.* New York: Van Nostrand Reinhold.

Hwang, S.-C., R.A. Larson, and V.L. Snoeyink. 1990. Reactions of Free Chlorine With Substituted Anilines in Aqueous Solution and on Granular Activated Carbon. *Water Res.*, 24(4):427–432.

Kalkhoff, S.J., D.W. Kolpin, E.M. Thurman, I. Ferrer, and D. Barceló. 1998. Degradation of Chloroacetanilide Herbicides: The Prevalence of Sulfonic and Oxanilic Acid Metabolites in Iowa Groundwater and Surface Waters. *Environ. Sci. Technol.*, 32(11):1738–1740.

Katz, E.L. 1986. Reactivity of Nitrogeneous and Other Organic Compounds With Aqueous Chlorine. *Bull. Environ. Contam. Toxicol.*, 36:715–722.

Kimmel, E.C., J.E. Casida, and L.O. Ruzo. 1986. Formamidine Insecticides and Chloroacetanilide Herbicides: Disubstituted Anilines and Nitrosobenzenes as Mammalian Metabolites and Bacterial Mutagens. *Jour. Agric. Food Chem.*, 34(2):157–161.

Kochany, J., and R.J. Maguire. 1994. Sunlight Photodegradation of Metolachlor in Water. *Jour. Agric. Food Chem.*, 42:406–412.

Koga, M., K. Kadokami, and R. Shinohara. 1992. Laboratory-Scale Ozonation of Water Contaminated With Trace Pesticides. *Water Sci. Technol.*, 26(9–11):2257–2260.

Kolpin, D.W., J.E. Barbash, and R.J. Gilliom. 1998. Occurrence of Pesticides in Shallow Groundwater of the United States: Initial Results From the National Water Quality Assessment Program. *Environ. Sci. Technol.*, 32:558–566.

Kolpin, D.W., J.E. Barbash, and R.J. Gilliom. 2000. Pesticides in Ground Water of the United States, 1992–1996. *Ground Water*, 38(6):858–863.

Kolpin, D.W., B.K. Nations, D.A. Goolsby, and E.M. Thurman. 1996. Acetochlor in the Hydrologic System in Midwestern United States, 1994. *Environ. Sci. Technol.*, 30:1459–1464.

Kolpin, D.W., E.M. Thurman, and D.A. Goolsby. 1996. Occurrence of Selected Pesticides and Their Metabolites in Near-Surface Aquifers of the Midwestern United States. *Environ. Sci. Technol.*, 30:335–340.

Kolpin, D.W., E.M. Thurman, and S.M. Linhart. 1998. The Environmental Occurrence of Herbicides: The Importance of Degradates in Groundwater. *Arch. Environ. Contam. Toxicol.*, 35:385–390.

Kolpin, D.W., E.M. Thurman, and S.M. Linhart. 2000. Finding Minimal Herbicide Concentrations in Ground Water? Try Looking for Their Degradates. *Sci. Total Environ.*, 248:115–122.

Kolpin, D.W., E.M. Thurman, and S.M. Linhart. 2001. Occurrence of Cyanazine Compounds in Groundwater: Degradates More Prevalent Than the Parent Compound. *Environ. Sci. Technol.*, 35:1217–1222.

Kross, B.C., A. Vergara, and L.E. Raue. 1992. Toxicity Assessment of Atrazine, Alachlor, and Carbofuran and Their Respective Metabolites Using Microtox. *Jour. Toxicol. Environ. Health*, 37:149–159.

Kruithof, J.C., R. Hopman, R.T. Meijers, and J.A.M.H. Hofman. 1994. Presence and Removal of Pesticides in Dutch Drinking Water Practice. *Water Supp.*, 12(SS 11):10–14.

Laue, H., J.A. Field, and A.M. Cook. 1996. Bacterial Desulfonation of the Ethanesulfonate Metabolite of the Chloroacetanilide Herbicide Metazachlor. *Environ. Sci. Technol.*, 30(4):1129–1132.

Liu, S.Y., A.J. Freyer, and J.-M. Bollag. 1991. Microbial Dechlorination of the Herbicide Metolachlor. *Jour. Agric. Food Chem.*, 39(3):631–636.

Liu, S.-Y., Z. Zheng, R. Zhang, and J.-M. Bollag. 1989. Sorption and Metabolism of Metolachlor by a Bacterial Community. *Appl. Environ. Microbiol.*, 55(3):733–740.

Loch, A.R., K.A. Lippa, D.L. Carlson, Y.P. Chin, S.J. Traina, and A.L. Roberts. 2002. Nucleophilic Aliphatic Substitution Reactions of Propachlor, Alachlor, and Metolachlor With Bisulfide (HS^-) and Polysulfides (S_n^{2-}). *Environ. Sci. Technol.*, 36(19):4065–4073.

Mathew, R., and S.U. Kahn. 1996. Photodegradation of Metolachlor in Water in the Presence of Soil Mineral and Organic Constituents. *Jour. Agric. Food Chem.*, 44:3996–4000.

Miltner, R.J., D.B. Baker, T.F. Speth, and C.A. Fronk. 1989. Treatment of Seasonal Pesticides in Surface Waters. *Jour. AWWA*, 81(1):43–52.

Nelieu, S., L. Kerhonas, and J. Einhorn. 2000. Degradation of Atrazine Into Ammeline by Combines Ozone/Hydrogen Peroxide Treatment in Water. *Environ. Sci. Technol.*, 34:430–437.

Nelson, G.B., and J.A. Ross. 1998. DNA Adduct Formation by the Pesticide Alachlor and Its Metabolite 2-Chloro-*N*-(2,6-Diethylphenyl) Acetamide (CDEPA). *Bull. Environ. Contam. Toxicol.*, 60:387–394.

Nesnow, S., S.C. Agarwal, W.T. Padgett, G.R. Lambert, P. Boone, and A.M. Richard. 1995. Synthesis and Characterization of Adducts of Alachlor and 2-Chloro-*N*-(2,6-Diethylphenyl)Acetamide with 2'-Deoxyguanosine, Thymidine, and Their 3'-Monophosphates. *Chem. Res. Toxicol.*, 8:209–217.

Osano, O., W. Admiral, and D. Otieno. 2002. Developmental Disorders in Embryos of the Frog *Xenopus Laevis* Induced by Chloroacetanilide Herbicides and Their Degradation Products. *Environ. Toxicol. Chem.*, 21(2):375–379.

Peñeula, G.A., and D. Barceló. 1996. Comparative Degradation Kinetics of Alachlor in Water by Photocatalysis With $FeCl_3$, TiO_2 and Photolysis, Studied by Solid-Phase Disk Extraction Followed by Gas Chromatographic Techniques. *Jour. Chromatogr., A*, 754:187–195.

Pereira, W.E., C.E. Rostad, and T.J. Leiker. 1990. Determination of Trace Levels of Herbicides and Their Degradation Products in Surface and Ground Waters by Gas Chromatography/Ion-Trap Mass Spectrometry. *Anal. Chim. Acta*, 228:69–75.

Phillips, P.J., G.R. Wall, E.M. Thurman, D.A. Eckhardt, and J. Vanhoesen. 1999. Metolachlor and Its Metabolites in Tile Drain and Stream Runoff in the Canajoharie Creek Watershed. *Environ. Sci. Technol.*, 33:3531–3537.

Pothuluri, J.V., F.E. Evans, D.R. Doerge, M.I. Churchwell, and C.E. Cerniglia. 1997. Metabolism of Metolachlor by the Fungus *Cunninghamella Elegans*. *Arch. Environ. Contam. Toxicol.*, 32:117–125.

Potter, T.L., and T.L. Carpenter. 1995. Occurrence of Alachlor Environmental Degradation Products in Groundwater. *Environ. Sci. Technol.*, 29:1557–1563.

Rakitsky, V.A., V.A. Koblyakov, and V.S. Turusov. 2000. Nongenotoxic (Epigenetic) Carcinogens: Pesticides as an Example. A Critical Review. *Teratog., Carcinog., and Mutag.*, 20:229–240.

Randtke, S.J., A.A. Witt, P.V. Adams, R.E. Pancake, and C.D. Adams. 1994. Occurrence and Control of Atrazine and Its Degradation Products in Drinking Water Supplies. In *Proc. of the AWWA Annual Conference*. Denver, Colo.: AWWA.

Sanyal, D., and G. Kulshrestha. 2002. Metabolism of Metolachlor by Fungal Cultures. *Jour. Agric. Food Chem.*, 50:499–505.

Shoemaker, J.A. 2002. Novel Chromatographic Separation and Carbon Solid-Phase Extraction of Acetanilide Herbicide Degradation Products. *Jour. AOAC Int.*, 85(6):1331–1337.

Snoeyink, V.L. 1990. Adsorption of Organic Compounds. In *Water Quality and Treatment: A Handbook of Community Water Supplies*. 4th ed. Edited by F.W. Pontius. New York: McGraw-Hill.

Somich, C.J., P.C. Kearney, M.T. Muldoon, and S. Elsasser. 1988. Enhanced Soil Degradation on Alachlor by Treatment With Ultraviolet Light and Ozone. *Jour. Agric. Food Chem.*, 36(6):1322–1326.

Tessier, D.M., and J.M. Clark. 1995. Quantitative Assessment of the Mutagenic Potential of Environmental Degradative Products of Alachlor. *Jour. Agric. Food Chem.*, 43:2504–2512.

Thurman, E.M., D.A. Goolsby, D.S. Aga, M.L. Pomes, and M.T. Meyer. 1996. Occurrence of Alachlor and Its Sulfonated Metabolite in Rivers and Reservoirs of the Midwestern United States: The Importance of Sulfonation in the Transport of Chloroacetamide Herbicides. *Environ. Sci. Technol.*, 30(2):569–574.

USGS (U.S. Geological Survey). 1998. Annual Use Maps [Online]. NAWQA Pesticide National Synthesis Project. Available: <http://ca.water.usgs.gov/pnsp/use92/>. [cited May 14, 2002]

White, M.C., J.D. Thompson, G.W. Harrington, and P.C. Singer. 1997. Evaluating Criteria for Enhanced Coagulation Compliance. *Jour. AWWA*, 89(5):64–77.

Wilson, R.I., and S.A. Mabury. 2000. Photodegradation of Metolachlor: Isolation, Identification and Quantification of Monochloroacetic Acid. *Jour. Agric. Food Chem.*, 48:944–950.

Winslow, S.D., B. Prakash, M.M. Domino, B.V. Pepich, and D.J. Munch. 2000. United States Environmental Protection Agency, National Exposure Research Laboratory. *EPA Method 526: Determination of Selected Semivolatile Organic Compounds in Drinking Water by Solid Phase Extraction and Capillary Column Gas Chromatography/Mass Spectrometry (GC/MS)* [Online]. Available: <http://www.epa.gov/safewater/methods/526.pdf>. [cited April 2002]

Winslow, S.D., B. Prakash, M.M. Domino, B.V. Pepich, and D.J. Munch. 2001a. Considerations Necessary in Gathering Occurrence Data for Selected Unstable Compounds in the USEPA Unregulated Contaminant Candidate List in USEPA Method 526. *Environ. Sci. Technol.*, 35:1851–1858.

Winslow, S.D., B.V. Pepich, M.V. Bassett, S.C. Wendelken, D.J. Munch, and J.L. Sinclair. 2001b. Microbial Inhibitors for USEPA Drinking Water Methods for the Determination of Organic Compounds. *Environ. Sci. Technol.*, 35:4103–4110.

Yao, C.C.D., and W.R. Haag. 1991. Rate Constants for Direct Reactions of Ozone With Several Drinking Water Contaminants. *Water Res.*, 25(7):761–773.

Zheng, H., and C. Ye. 2001. Identification of UV Products and Hydrolysis Products of Butachlor by Mass Spectrometry. *Environ. Sci. Technol.*, 35:2889–2895.

ABBREVIATIONS

AI	active ingredient
APHA	American Public Health Association
ARP	acetochlor registration partnership
AWWA	American Water Works Association
AwwaRF	Awwa Research Foundation
°C	degrees Celsius
CCL	Contaminant Candidate List
C_e	equilibrium solution concentration
DCM	dichloromethane
DNA	deoxyribonucleic acid
DOC	dissolved organic carbon
DOM	dissolved organic matter
EI	electron ionization
ELISA	enzyme-linked immunosorbent assay
EPA	Environmental Protection Agency
eq	equivalents
ESA	ethane sulfonic acid
eV	electron volt
FT-NMR	Fourier transform nuclear magnetic resonance
g	gram
GAC	granular activated carbon
GC	gas chromatography
GC/MS	gas chromatography/mass spectrometry
g/L	grams per liter
GSH	glutathione
^{1}H-NMR	proton nuclear magnetic resonance
HPLC-DAD	high-pressure liquid chromatography-diode array detection
hrs	hours
ID	inner diameter
K	Freundlich isotherm coefficient
K_{ow}	octanol water partition coefficient
L	liter
LVI	large volume injection

m	meter
M	molar
MCL	maximum contaminant level
MCLG	maximum contaminant level goal
MDL	method detection limit
MGD	millions of gallons per day
mg/L	milligrams per liter
MHz	megahertz
min	minute
mL	milliliter
mM	millimolar
MNNG	*N*-methyl-*N'*-nitro-*N*-nitrosoguanidine
MRL	minimum reporting limit
MS	mass spectrometer
m/z	mass to charge ratio
μg/L	micrograms per liter
μL	microliter
μm	micron
N	normal
ng/L	nanograms per liter
nm	nanometer
NOM	natural organic matter
Nt	not tested
OA	oxanilic acid
PAC	powdered activated carbon
pCI	positive chemical ionization
PTV	programmed temperature vaporizer
q_e	sorption capacity
rpm	revolutions per minute
sec	second
SIM	selected ion monitoring
SPE	solid-phase extraction
TMS	tetramethylsilane
TOC	total organic carbon
U.S.	United States
USEPA	United States Environmental Protection Agency
USGS	United States Geological Survey
UV	ultraviolet

Printed in the United States
56911LVS00002B/515-528

9 781583 214473